AIR QUALITY: ISSUES AND OUTLOOK

AIR QUALITY: ISSUES AND OUTLOOK

LAWRENCE S. TOOKER
(EDITOR)

Nova Science Publishers, Inc.
New York

Production Coordinator: Tatiana Shohov
Coordinating Editor: Tatiana Shohov
Senior Production Editors: Susan Boriotti and Donna Dennis
Production Editors: Marius Andronie and Rusudan Razmadze
Office Manager: Annette Hellinger
Graphics: Levani Chlaidze and Magdalena Nuñez
Editorial Production: Alexandra Columbus, Maya Columbus, Robert Brower, Vladimir Klestov, Matthew Kozlowski and Lorna Loperfido
Circulation: Luis Aviles, Raymond Davis, Cathy DeGregory, Melissa Diaz, Ave Maria Gonzalez, Marlene Nuñez, Jeannie Pappas, Vera Popovic and Frankie Punger
Communications and Acquisitions: Serge P. Shohov

Library of Congress Cataloging-in-Publication Data
Available Upon Request

ISBN: 1-59454-075-6.

400 Oser Ave, Suite 1600
Hauppauge, New York 11788-3619
Tele. 631-231-7269 Fax 631-231-8175
e-mail: Novascience@earthlink.net
Web Site: http://www.novapublishers.com

Printed in the United States of America

CONTENTS

PREFACE

Everyone in the modern diversity drenched, multimedia, multitasking, multi-channel society seems to have their own agenda, activity and polarized point of view. One of the few things which unites them all is the necessity to breath air, and the less harmful, the better. The air, on the other hand, although presumably not opposed to providing the agenda gobblers with that which is essential to them, is being bombarded with smoke, auto exhaust fumes, and every kind of pollutant known to exist. This book presents the latest information on these crucial issues including the usual feeble legislation being offered up, analyses of air quality and electricity, and the elusive standards necessary to maintain a decent level of relativity clean air.

With the prospect of new layers of complexity being added to air pollution controls and with electricity restructuring putting a premium on economic efficiency, interest is being expressed in finding mechanisms to achieve health and environmental goals in simpler, more cost-effective ways. The electric utility industry is a major source of air pollution, particularly sulfur dioxide (SO_2), nitrogen oxides (NO_x), and Mercury (Hg), as well as suspected greenhouse gases, particularly carbon dioxide (CO_2). At issue in chapter one is whether a new approach to environmental protection could achieve the nation's air quality goals more cost-effectively than the current system.

As of January 22, 2001, the EPA has declared 11 states and the District of Columbia as failing to submit revised SIPs required under the Ozone Transport Rule; the EPA has approved four section 126 petitions; and one of the NSR lawsuits has resulted in a consent decree, and two others have been settled in principle. Legislative activity focuses on developing a multi-pollutant strategy as an alternative to these piecemeal initiatives. In June 2002, the Bush Administration recommended new rulemaking be commenced on the definition of "routine maintenance." Legislative activity focuses on multi-pollutant strategies as an alternative to these piecemeal initiatives. In June 2002, the Senate Environment and Public Works Committee reported out S. 556, a comprehensive, multi-pollutant reduction bill. Chapter two focuses on this issue.

In addition to changing New Source Review (NSR), the Administration has asked Congress to modify Clean Air Act requirements for power plants by enacting “Clear Skies” or “multi-pollutant” legislation. Another holdover issue concerns regulation of the gasoline additive MTBE. Other clean air issues that might be considered in the 108th Congress are the conformity of metropolitan area transportation plans with the Clean Air Act, and whether to modify the Act's requirements for areas that have not met deadlines for attainment of the ozone air quality standard. They are all discussed in chapter three.

In the context of federal and state proposals to restructure the electric utility industry, chapter four analyzes forces and policies affecting utility generation that may have consequences for emissions of air pollutants and or greenhouse gases. Key concerns are potential increases in nitrogen oxide emissions, raising questions about the effectiveness of the Clean Air Act to regulate a restructured industry, and in carbon dioxide emissions, which are not currently regulated but could be if the U.S. ratifies the Kyoto Agreement. These issues may be raised in the context of electricity restructuring legislation.

On February 27, 2001, the Supreme Court handed down its decision in *Whitman v. American Trucking Associations*, a challenge to EPA's promulgation in 1997 of revised national ambient air quality standards for ozone and particulates under the Clean Air Act. The Court charged the agency with developing a reasonable interpretation accommodating both subparts. Such accommodation is likely to provide a difficult task, however, and almost certainly once adopted will generate further legal challenges. Chapter five delves into this issue.

On November 3, 1999, the Justice Department filed seven lawsuits against electric utilities in the Midwest and south charging them with violations of the New Source Review (NSR) requirements of the Clean Air Act (CAA). In addition, the Environmental Protection Agency (EPA) issued an administrative order against the Tennessee Valley Authority (TVA), alleging similar violations. Chapter six reviews the NSR requirements and the suits; it identifies the facilities involved; and it considers issues of possible congressional interest that could arise from complications because of interactions of NSR with other provisions of the CAA and because of changing practices in the electric generating industry.

Chapter seven provides a side-by-side comparison of the major provisions of environmentally related legislation introduced with respect to electric restructuring.

EPA is currently engaged in a series of regulatory actions to address the transport of ozone pollution in the eastern United States. Chapter eight reviews this situation with respect to the EPA's Ozone Transport Rule and other activities.

The Environmental protection Agency (EPA) has completed lengthy negotiations with numerous states and vehicle manufacturers to establish a voluntary National Low Emission Vehicle (LEV) Program. Manufacturers have agreed to produce cleaner vehicles for sale in eight Northeastern states and the District of Columbia beginning in model year 2000 and nationwide in model year 2001. The program uses the same low emission standards and testing procedures already adopted in California, New York, Massachusetts, Vermont, and Maine. Chapter nine provides background information on federal emission standards for motor vehicles under the Clean Air Act and stricter standards originally developed to address the severity of air quality problems in California, explains the low emission standards and flexible compliance mechanisms to which states and manufacturers have voluntarily agreed under the National LEV Program, discusses EPA's estimates of the program's air quality benefits and costs, and examines regulatory issues related to its implementation including sulfur levels in gasoline and the relative stringency of emission standards for light trucks.

Emissions trading is a market-based mechanism used to lower the costs of complying with air quality regulations. It allows sources facing high pollution control costs to meet their emission limits by purchasing excess reductions from other sources that can afford to lower their emissions further than federal or state regulations require. Chapter ten serves as a primer on the concept of emissions trading.

Chapter eleven provides background information on the regulation of vehicle emissions in the United States, analyzes key elements of the National LEV program and EPA's Tier 2 proposal, summarizes major views on the proposal that have been expressed by the automobile industry, the oil refining industry, and some environmental organizations, and discusses relevant legislative activity in the 106th Congress.

The Administration has announced that the recommendations of the Ozone Transport Assessment Group (OTAG) will provide the basis for its rule on regional compliance strategies to assist nonattainment areas to achieve the ozone National Ambient Air Quality Standard (NAAQS). This issue is discussed in chapter twelve.

The decisions by the Administrator of the Environmental Protection Agency (EPA) in 1997 to revise the National Ambient Air Quality Standards for ozone and particulate matter refocused attention on the criteria and the process by which these decisions are made. Tracing the steps of the decision process, chapter thirteen identifies the statutory criteria established by the Congress and summarizes the administrative procedures the Agency follows in setting these standards--and in reviewing them every five years.

Chapter fourteen provides background information concerning methyl tertiary butyl ether (MTBE), an additive used to produce cleaner burning gasoline, and legislation that might affect its use. Controversy has surrounded the use of MTBE, most recently because of concern over contamination of drinking water supplies by leaking gasoline storage tanks, pipelines, and marine engines. It summarizes information concerning the environmental impacts of the additive's use and potential regulatory and legislative options.

Most sport utility vehicles and mini-vans are classified as light trucks and are regulated less stringently than passenger cars under two major laws--the Energy Policy and Conservation Act for fuel economy standards, and the Clean Air Act for emissions standards. Chapter fifteen discusses the discrepancy between emissions and fuel economy standards for passenger cars and light trucks, how that discrepancy is changing, and legislative activity.

Chapter 1

AIR QUALITY: MULTI-POLLUTANT LEGISLATION*

Larry B. Parker and John E. Blodgett

INTRODUCTION

Electric utility generating facilities are a major source of air pollution. The combustion of fossil fuels (petroleum, natural gas, and coal), which account for 67% of U.S. electricity generation, results in the emission of a stream of gases. These gases include several pollutants that directly pose risks to human health and welfare, including particulate matter (PM),[1] sulfur dioxides (SO_2), nitrogen oxides (NO_x), and mercury (Hg). Particulate matter, sulfur dioxide and NO_x are currently regulated under the Clean Air Act (CAA), and the Environmental Protection Agency (EPA) has announced its intention to regulate mercury. Other gases may pose indirect risks, notably carbon dioxide (CO_2), which may contribute to global warming.[2] Table 1 provides estimates of SO_2, NO_x, and CO_2 emissions from electric generating facilities. Annual emissions of Hg from utility facilities are more uncertain; current estimates indicate about 48 tons. Utilities are subject to an array of environmental regulations, which affect in different ways both the cost of operating existing generating facilities and of constructing new ones.

* Excerpted from CRS Report RL31326. Updated October 22, 2002.

[1] Particulate matter is regulated depending on the particle size; current regulations address particles less than 10 microns in diameter (PM_{10}); EPA has promulgated regulations for particles less than 2.5 microns in diameter ($PM_{2.5}$) which have not been implemented. SO_2 and NO_x emissions could be affected by regulations of $PM_{2.5}$. Current concerns about emissions from fossil-fuel electric generating plants do not explicitly address PM, but could indirectly do so through attention to SO_2 and NO_x.

[2] Steam-electric utilities produce minor amounts of volatile organic compounds (VOCs), carbon monoxide (CO), and lead — on the order of 2% or less of all sources.

Table 1: Emissions From U.S. Fossil-fuel Electric Generating Plants (thousands of short tons)

Emissions	1994	1995	1996	1997	1998	1999
SO_2	14,211	11,437	12,053	12,317	12,432	11,968
NO_x	6,790	6,737	6,996	7,227	7,221	7,051
CO_2	1,986,079	1,995,471	2,065,339	2,142,118	2,209,286	2,191,576

Source: Energy Information Administration, *Electric Power Annual 1999*, vol. II, p. 40

The evolution of air pollution controls over time and as a result of growing scientific understanding of health and environmental impacts has led to a multilayered and interlocking patchwork of controls. Moreover, additional controls are in the process of development, particularly with respect to NO_x as a precursor to ozone, to both NO_x and SO_2 as contributors to $PM_{2.5}$, and to Hg as a toxic air pollutant. Also, under the United Nations Framework Convention on Climate Change (UNFCCC), the United States agreed to voluntary limits on CO_2 emissions. The current Bush Administration has rejected the Kyoto Protocol, which would impose mandatory limits, in favor of a voluntary reduction program. Thus, mandatory federal CO_2 controls in the United States appear unlikely in the near term. For many years the complexity of the air quality control regime has caused some observers to call for a simplified approach. Now, with the potential both for additional control programs on SO_2 and NO_x and for new controls directed at Hg and CO_2 intersecting with the technological and policy changes affecting the electric utility industry, such calls for simplification have become more numerous and insistent. One focus of this effort is the "multi-pollutant" or "four pollutant" approach. This approach involves a mix of regulatory and economic mechanisms that would apply to utility emissions of up to four pollutants in various proposals – SO_2, NO_x, Hg, and CO_2. The objective would be to balance the environmental goal of effective controls across the pollutants covered with the industry goal of a stable regulatory regime for a period of years.[3]

PROPOSED LEGISLATION AND LEGISLATIVE ACTION IN THE 107TH CONGRESS

Currently, eight bills have been introduced in the 107th Congress to reduce emissions by increasing pollution controls on electric generating facilities. They are summarized in Appendix 1. All of the bills control at least NO_x and SO_2; others include CO_2 and Hg. All of these bills involve some form of emission caps, and most include a tradeable credit program to implement that cap.

Two bills, H.R. 25 and S. 588, are companion legislation focused on SO_2 and NO_x, with a mandate to EPA to regulate Hg by 2005. The bills build on the SO_2 allowance trading scheme contained in title IV of the 1990 Clean Air Act Amendments (CAAA); under this program utilities are given a specific allocation of permitted emissions (called allowances) and may

[3] Larry Parker and John Blodgett, *Electricity Generation and Air Quality: Multi-pollutant Strategies,* CRS Report RL30878, March 13, 2001.

choose to use those allowances at their own facilities, or, if they do not use their full quota, to bank them for future use or to sell them to other utilities needing additional allowances.

Two other bills, H.R. 1256 and S. 556 as originally introduced, are the same except for minor wording differences. The bills would control SO_2, NO_x, Hg, and CO_2, but provide considerable flexibility to EPA in developing implementation strategies. Specifically, EPA implementing regulations are required to allocate reductions equitably and may consider market-oriented mechanisms, except for Hg.

The fifth bill introduced, H.R. 1335, controls SO_2, NO_x, Hg, and CO_2, but with individual unit-by-unit requirements for SO_2 and NO_x based on output-based emission rates and average 1998-2000 fuel consumption[4]; a percentage reduction requirement for Hg; and an allowance-based system for CO_2.

The next two bills introduced, H.R. 5266 and S. 2815, are the Administration's proposal to control SO_2, NO_x, and Hg. It is discussed in the next section along with the Administration's voluntary CO_2 initiative.

The final bill, S. 3135, controls SO_2, NO_x, Hg, and CO_2 with an allowance-based trading system that includes a new source reserve to allocate allowances to newly constructed sources. The bill includes provisions that would revise Clean Air Act (CAA) provisions with respect to the New Source Review (NSR) trigger for modified or reconstructed boilers, and include a cost threshold for the Lowest Available Emission Rate (LAER) requirement in non-attainment areas. In addition, S. 3135 would eliminate the CAA's offset requirement for new electric generating facilities constructed in non-attainment areas beginning in 2008.

On June 27, 2002, the Senate Environment and Public Works Committee reported out an amended version of S. 556. As indicated in Appendix 1, in comparison with the bill as introduced, the amended S. 566's compliance deadlines for its reduction requirements have been extended one year to 2008. In addition, the reported bill elaborates on provisions with respect to excess emissions penalties and protections for possible local ambient air impacts. In particular, the reported bill has detailed provisions for allocating SO_2, NO_x and CO_2 allowances to various economic sectors and interests. In most cases, these interests (or their trustees in the case of households and dislocated workers and communities) would auction off (or otherwise sell) their allowances to the affected utilities, and use the collected funds for their designated purposes.

THE BUSH ADMINISTRATION'S PROPOSALS

In February 2002, the Bush Administration announced two air quality proposals.[5] The first would amend the Clean Air Act to place emission caps on electric utility emissions of SO_2, NO_x, and Hg. Implemented through a tradeable allowance program, the emissions caps would be imposed in two phases: 2010 (2008 in the case of NO_x) and 2018. It was introduced as part of a complete rewrite of Title IV of the Clean Air Act on July 26 as H.R. 5266. It was introduced in the Senate on July 29 as S. 2815.

[4] Plants built after 2000 appear to be in an anomalous position; as their historical fuel consumption would be zero, the bill would seem to imply they can emit no SO_2 or NO_x.

[5] Papers outlining the Administration's proposals are available from the White House web site: [http://www.whitehouse.gov/news/releases/2002/02/clearskies.html] for the three pollutant proposal and, for the climate change initiative: [http://www.whitehouse.gov/news/releases/2002/02/climatechange.html].

As indicated in Appendix 1, H.R. 5266/S. 2815's NO_x cap is significantly less stringent for 2008 than the requirements for 2005 or 2007 in most of the bills discussed above, and remains less stringent even through the second phase beginning in 2018. The situation is similar for SO_2, except that its 2010 cap is similar to the H.R. 25/S. 588 cap for 2007 (and S. 3135 for 2008), and its 2018 cap is similar to the H.R. 1335 cap for 2005. On mercury, H.R. 5266/S. 2815's caps are similar to those in S. 3135 but come into effect later. H.R. 5266/S. 2815's 2018 Hg emissions goal allows about 3 times more emissions than those allowed by S. 556, H.R. 1256, and H.R. 1335 for 2005, 2007, or 2008. (It is not really possible to compare H.R. 25/S. 588, which do not specify an Hg emissions goal, leaving regulation up to EPA, except that the bills set a regulatory deadline of 2005 while the Administration proposal would begin regulating Hg in 2010.)

In addition to the emissions caps, H.R. 5266/S.2815 would substantially modify or eliminate several provisions in the Clean Air Act with respect to electric generating facilities. The New Source Performance Standards (Section 111) would be eliminated and replaced with standards for SO_2, NO_x, particulate matter, and Hg for new sources, and modified sources that opt to comply with them in lieu of Best Available Control Technology (BACT) determinations under Prevention of Significant Deterioration (PSD) provisions (CAA, Part C) or Lowest Achievable Emissions Rate (LAER) determination under non-attainment provisions (CAA, Part D). Compliance with these provisions exempts such facilities from New Source Review (NSR), PSD-BACT requirements, visibility (Best Available Retrofit Technology) BART requirements, and non-attainment LAER requirements. The exemption does not apply to PSD-BACT requirements if facilities are within 50 km of a PSD Class 1 area.

Existing sources can also receive these exemptions if they agree to meet a particulate matter standard specified in the bill along with good combustion practices to minimize carbon monoxide emissions within 3 years of enactment.

Other exemptions provided by H.R. 5266/S. 2815 include an exemption for steam electric generating facilities from regulation under Section 112 of the CAA, and relief from enforcement of any Section 126 petition before 2012. As discussed earlier, S. 3135 is the only other bill that contains any regulatory relief provisions.

The second Bush administration proposal (for which no legislation has been introduced) initiates a new voluntary greenhouse gas reduction program, similar to ones introduced by the earlier Bush and Clinton Administrations.[6] Developed in response to the U.S. ratification of the 1992 UNFCCC, these previous plans projected U.S. compliance, or near compliance, with the UNFCCC goal of stabilizing greenhouse gas emissions at their 1990 levels by the year 2000 through voluntary measures. The new proposal introduced by the Bush Administration does not make that claim, only projecting a 100 million metric ton reduction in emissions from what would occur otherwise in the year 2012. Instead, the plan focuses on improving the carbon efficiency of the economy, reducing current emissions of 183 metric tons per million dollars of GDP to 151 metric tons per million dollars of GDP in 2012. It proposes several voluntary initiatives, along with increased spending and tax incentives, to achieve this goal. However, the Administration states that three-quarters of the projected reduction would be achieved through current efforts underway, not by the new initiatives.

[6] For a discussion of those previous plans, see Larry Parker and John Blodgett, *Climate Change Action Plans*, CRS Report 94-404 ENR, May 9, 1994. (archived, available from the author)

It is difficult to compare the Administration's proposal to the bills that have been introduced, as it is a voluntary, not mandatory program. S. 556, H.R. 1256, and H.R. 1335 set their caps at 1990 levels. S. 3135 sets its final cap at 2001 levels. In contrast, the Administration's proposal contains no cap or other limits on emissions. Rather, following general climate change approaches of the earlier Bush Administration and the Clinton Administration, the Administration's proposal relies on various voluntary programs and incentives to encourage reductions in greenhouse gases from diverse sources, including CO_2 from electric generation.

Based on the estimate provided by the Administration's climate change proposal, and using the 2002 *Climate Action Report*[7] (CAR) for projections to 2010, table 2 presents estimates of U.S. greenhouse gas emissions in 2010, assuming the Administration's voluntary program reaches its goals.[8] This should not be taken as a given, as neither the former Bush Administration's program nor the Clinton Administration's program achieved their stated goals. Thus, in one sense, comparing a mandatory reduction program, such as those proposed in legislation, with the Administration's voluntary program is comparing apples to oranges. The first is legally binding, the second is an exhortation.

While S. 556, H.R. 1256, and H.R. 1335 address only electric utility emissions, their mandated reductions would result in lower greenhouse gas emissions in 2010 than those projected to occur under the Administration's initiative that includes all sources of all greenhouse gases.[9] Likewise, S. 3135 CO_2 control requirement, while less stringent than S. 556, H.R. 1256, and H.R. 1335, is also estimated to achieve more reductions than the Administration's initiative. However, neither S. 556, H.R. 1256, H.R. 1335, S. 3135 nor the Administration's initiative would be sufficient to bring U.S. emissions anywhere near the level committed to at Rio with the UNFCCC.

Indeed, discussion in the CAR suggests that a high economic growth scenario would significantly increase energy use and related carbon emissions. For example, under a high economic growth scenario, greenhouse emissions in 2010 would increase 37.7% above those in 1990, based on energy growth alone. This increase would represent an additional 53 million metric tons of emissions.[10] However, S. 556, H.R. 1256, and H.R. 1335 would cap emissions from increased electricity generation at 1990 levels, which would reduce the 53 million metric tons by 16 million metric tons, or 30% of the high growth increase. The Administration's initiative is voluntary, and contains no such mandatory caps on emissions growth.

[7] This is the U.S. report to the UNFCCC Secretariat on U.S. emissions and measures taken to reduce them. The *Climate Action Report— 2002*, available at: http://www.epa.gov/globalwarming/publications/car/index.html

[8] For a discussion of emission projections and trends, see John Blodgett and Larry Parker, *Global Climate Change: U.S. Greenhouse Gas Emissions – Status, Trends, and Projections*, CRS Report 9-235 ENR (February 28, 2002).

[9] The assessment assumes that the Administration's proposal actually achieves its goal in 2010, rather than 2012.

[10] Energy Information Administration, *Annual Energy Outlook 2000*, Washington D.C., DOE/EIA-0383 (2002), December 2001. p. 177.

Table 2: Comparison of Administration Voluntary Program with Proposed Legislation

	Percentage Change v. Business as Usual (2010)	**Percentage Change v. 1990 levels per UNFCCC**
S. 556, H.R. 1256, H.R. 1335	-7.5%	+24.2%
S. 3135*	-5.1%	+27.5%
Administration Voluntary Program*	-4.4 to -4.5%	+28.3%
Business as Usual	0	+34.4%

*Assumes requirement (S. 3135) or goal (Administration Voluntary Program) is achieved in 2010, rather than 2012.
Source: CRS calculations based on projections contained in 2002 CAR.

Appendix 1: Comparison of Multi-Pollutant Control Proposals

Provisions	H.R. 25 (Sweeney)/S. 588 (Schumer)	S. 556 (as reported) (Jeffords)	H.R. 1256 (Waxman)/S. 556 (as introduced) (Jeffords)	H.R. 1335 (Allen)	H.R. 5266/S. 2815 (Administration's Proposals)	S. 3135 (Carper)
Emissions Cap on NO_x	estimated at 1.5 million tons in 2007 with interim reductions	estimated at 1.5 million tons in 2008	estimated at 1.5 million tons in 2007	estimated at 1.6 million tons in 2005, declining with plant retirements	2.1 million tons in 2008, declining to 1.7 million tons in 2018	1.87 million tons in 2008, declining to 1.7 million tons in 2012
Emissions Cap on SO_2	4.45 million tons in 2007	2.25 million tons in 2008	2.23 million tons in 2007	estimated at 3.2 million tons in 2005, declining with plant retirements	4.5 million tons in 2010, declining to 3.0 million tons in 2018	4.5 million tons in 2008, declining to 2.25 million tons in 2015
Emission Cap on CO_2	not covered	estimated at 2.05 billion tons in 2008	estimated at 1.914 billion tons in 2007	1.914 billion tons in 2005	none, program is voluntary	estimated at 2.6 billion tons in 2008, declining to an estimated 2.3 billion tons in 2012
Emissions Cap on Mercury	EPA to regulate by 2005	estimated at 5 tons in 2008	estimated at about 4-5 tons in 2007	estimated at about 4-5 tons in 2005	26 tons in 2010, declining to 15 tons in 2018	24 tons in 2008, declining to 5-16 tons by 2012 according to an EPA determination
Scope	48 contiguous states and DC	50 states and DC	50 states and DC	50 states and DC	50 states, DC, and territories	50 states and DC

Provisions	H.R. 25 (Sweeney)/S. 588 (Schumer)	S. 556 (as reported) (Jeffords)	H.R. 1256 (Waxman)/S. 556 (as introduced) (Jeffords)	H.R. 1335 (Allen)	H.R. 5266/S. 2815 (Administration's Proposals)	S. 3135 (Carper)
Affected Units	electric generating facilities 25 Mw or greater	electric generating facilities 15 Mw or greater	electric generating facilities 15 Mw or greater	electric generating facilities 15 Mw or greater (50 Mw for CO_2)	for existing SO_2, NO_x, and Hg, electric generating facilities 25 Mw or greater; no size minimum on new facilities; voluntary CO_2 program is economy-wide	fossil fuel-fired electric generating facilities greater than 25 Mw (coal-fired facilities in the case of Hg controls)
Penalties for non-compliance	NO_x: \$6,000 per excess ton plus one-for-one offset from future emission allocations SO_2: same as CAA, title IV	NO_x and SO_2: same as CAA, title IV except excess emission penalty is three times the average market price for allowances CO_2: three times the average market price per excess metric ton of CO_2 emissions Hg: three times the	determined by EPA	NO_x, SO_2, Hg: no special penalties specified – CAA penalties would apply CO_2: \$100 per ton plus one-for-one offset from future emission allocations	NO_x, SO_2, Hg: reduces the excess emissions penalties under CAA, title IV to the lowest EPA auction price for allowances plus one-for-one offset from future emission allocations CO_2: none – voluntary program	NO_x: \$5,000 per ton plus one-for-one offset from future emission allocations SO_2: same as CAA, title IV Hg: \$10,000 per lb. plus one-for-one offset from future emission allocations CO_2: \$100 per ton

Provisions	H.R. 25 (Sweeney)/S. 588 (Schumer)	S. 556 (as reported) (Jeffords)	H.R. 1256 (Waxman)/S. 556 (as introduced) (Jeffords)	H.R. 1335 (Allen)	H.R. 5266/S. 2815 (Administration's Proposals)	S. 3135 (Carper)
		average Hg control costs per gram of excess emission				plus one-for-one offset from future emission allocations
Special Provisions	NO_x allowance value halved during ozone season; reserve of allowances for new sources	all powerplants 30-years or older must meet current New Source Performance Standard (NSPS) requirements SO_2 cap divided by region (West and East) other provisions to protect local air quality	all powerplants 30-years or older must meet current New Source Performance Standard (NSPS) requirements	permanent CO_2 and NO_x reductions through plant retirements should be credited in any future climate change implementation program enacted by Congress	new performance standards for new sources replace current NSPS for new sources. Compliance exempts such facilities from New Source Review (NSR), PSD-BACT requirements, visibility BART requirements, and non-attainment LAER requirements. The exemption does not apply to PSDBACT requirements if	tonnage limitations in effect for 20 years; EPA to reassess after 15 years CO_2 limitations include provisions providing CO_2 allowances for renewable energy, sequestration, and other greenhouse gas emission reduction projects revises the trigger for New Source Review (NSR) to modifications that exceed 50% of new

Provisions	H.R. 25 (Sweeney)/S. 588 (Schumer)	S. 556 (as reported) (Jeffords)	H.R. 1256 (Waxman)/S. 556 (as introduced) (Jeffords)	H.R. 1335 (Allen)	H.R. 5266/S. 2815 (Administration's Proposals)	S. 3135 (Carper)
					facilities is within 50 Km of Class 1 area. Existing sources can opt in by meeting a particulate standard Exempts utility units from regulation under CAA, Section 112 Prevents EPA from	construction costs; and Lowest Achievable Emission Rate (LAER) to include a cost threshold; in 2008, eliminates the offset requirement in non-attainment areas for new electric generating facilities
Implementation Strategy	tradeable allowance system	tradeable allowance system for SO_2 (restricted between East and West regions), NO_x and CO_2. Allowances allocated to various sectors and interests, including households, dislocated workers	to be determined by EPA — market mechanisms permitted (except for Hg)	unit-by-unit compliance with SO_2, NO_x, Hg provisions; tradeable allowance system for CO_2	tradeable allowance system for SO_2, NO_x, and Hg. Allocation formulas in the bill initially provide most allowances to affected sources free, with a small percentage sold at auction. Over time, an increasing	tradeable allowance system for SO_2, NO_x, Hg, and CO_2. Allocations formulas for NO_x, Hg, and CO_2 based on generating efficiency; SO_2 allocations based on current CAA, title IV provisions.

Provisions	H.R. 25 (Sweeney)/S. 588 (Schumer)	S. 556 (as reported) (Jeffords)	H.R. 1256 (Waxman)/S. 556 (as introduced) (Jeffords)	H.R. 1335 (Allen)	H.R. 5266/S. 2815 (Administration's Proposals)	S. 3135 (Carper)
		and communities, electricity intensive industries, affected utilities, energy efficiency and renewable energy activities, and sequestration activities Hg compliance on a unit-by-unit basis			percentage of the allocation is sold at auction with affected sources receiving fewer allowances free	Allocations formulas for all four pollutants include a new source reserve to provide allowances to newly constructed sources

Unless otherwise noted, estimates by CRS using Department of Energy and Environmental Protection Agency data

Chapter 2

Air Quality and Electricity: Initiatives to Increase Pollution Controls[*]

Larry B. Parker and John E. Blodgett

Background

Since the mid-1990s, the U.S. Environmental Protection Agency (EPA) has initiated actions that have resulted in regulatory mandates and enforcement actions that would, if implemented, substantially reduce air pollutants (particularly nitrogen oxides – NO_x) emitted by some electric generating facilities. An Ozone Transport Assessment Group (OTAG), formed by EPA in May 1995, laid the groundwork for the regulatory initiatives; it directly led to the Ozone Transport Rule (also called the NO_x SIP Call). In a supplementary action, 12 states petitioned EPA under Section 126 of the CAA, concerning interstate pollution, alleging that NO_x originating in upwind states prevented their attainment of ozone standards. An EPA Office of Enforcement & Compliance Assurance audit of New Source Review (NSR) applications required under provisions of the Clean Air Act (CAA) that began in late 1996 was the precursor to the enforcement initiative; it led in November 1999 to lawsuits against seven utilities in the Midwest and South and an administrative order against the Tennessee Valley Authority alleging violations of NSR requirements of the CAA.[1]

The first two initiatives, the Ozone Transport Rule and the Section 126 petitions, are related to each other substantively.[2] These initiatives would further control NO_x to assist states in the Northeast in meeting the existing, statutory 1-hour ozone National Ambient Air

* Excerpted from CRS Report RS20553. Updated October 25, 2002.

[1] For more details on the Ozone Transport Rule and Section 126 petitions, see CRS Report 98-236, *Air Quality: EPA's Ozone Transport Rule, OTAG, and Section 126 Petitions – A Hazy Situation?*, updated March 9, 2001. For more details on EPA's NSR action, see CRS Report RL30432, *Air Quality and Electricity: Enforcing New Source Review*.

[2] Until recently they were also linked administratively and procedurally through ties to another regulatory action by EPA – the 1997 promulgation of a new 8-hour National Ambient Air Quality Standard (NAAQS) for ozone. That linkage was broken by EPA when litigation temporarily halted implementation of the 8-hour standard. Some observers include the 8-hour standard as a related fourth initiative, but this report includes it only to the extent it is related to the NO_x SIP Call, the Section 126 petitions, and the enforcement action.

Quality Standard (NAAQS). The Ozone Transport Rule includes all or part of 19 eastern states and the District of Columbia. Based on the eight petitions EPA has ruled on, EPA's Section 126 determinations would involve a subset of the NO_x SIP Call's 19 states – 12 states and the District of Columbia.

The enforcement initiative is not legally or procedurally related to the above initiatives; however, the NSR enforcement action by EPA has substantive associations with them in that NO_x is a primary (but not sole) focus, and many of the utilities named as defendants in these cases would also have to reduce emissions under the NO_x SIP Call and Section 126 determinations. Unlike the other actions, the NSR action does not involve new regulatory action, but enforcement of existing law and regulations. As such, it is handled by the EPA's Office of Enforcement & Compliance Assurance, not a regulatory office, and involves other pollutants electric generators emit besides NO_x (specifically sulfur dioxide (SO_2) and particulates).

What is the Focus?

The primary focus of the regulatory initiatives and a primary effect of EPA's enforcement action is to reduce NO_x emissions in the eastern part of the United States. The environmental purpose for doing so is to reduce the interstate transportation of this ozone precursor, thus assisting localities along the eastern seaboard in attaining the ozone NAAQS. The actions would also mitigate acid rain. The initiatives and enforcement action by EPA focus on coal-fired electric generating facilities both because they are major sources of emissions – in 1997 they emitted 24% of the country's NO_x (and also 62% of its SO_2, 31% of its carbon dioxide (CO_2), and approximately one-third of the country's mercury (Hg)) – and because they represent the most cost-effective sources of large emission reductions for NO_x and SO_2.

In the case of the Section 126 determinations and the NSR enforcement action, coalfired powerplants are explicitly targeted for emissions reductions. In the case of the NO_x SIP Call, EPA cannot explicitly target sources (that is the responsibility of each affected state), but the allocation scheme used by EPA to determine the allowable emissions budget for individual states is based primarily on substantial reductions from coal-fired powerplants. In general, the initiatives identified here would require affected powerplants to reduce their NO_x emissions by about 75%-85%. Although the Section 126 determinations and the NSR enforcement action target individual sources, EPA provides flexibility for utilities to achieve the mandated reduction by means other than simply installing NO_x control equipment on affected units. As indicated by EPA's NSR settlement with Tampa Electric discussed later, the consent decree involves several different NO_x control strategies to reduce NO_x emissions by over 85%, as well as controls to reduce SO_2 emissions by almost 80%, by the year 2010.

Figure 1 indicates the states affected by the initiatives identified here. In line with the initiatives' focus on coal-fired electric generating facilities, the Midwest is the primary location of affected powerplants. Five states – Indiana, Kentucky, North Carolina, Ohio, and West Virginia – would be affected by all three initiatives. In contrast, Mississippi and Florida have utilities targeted only under the NSR enforcement initiatives; Missouri, Connecticut, Rhode Island, and Massachusetts are targeted only under the Ozone Transport Rule. The other states have utilities targeted under the Ozone Transport Rule and either a Section 126 determination or NSR enforcement.

States Potentially Affected by EPA's Recent Regulatory and Enforcement Actions

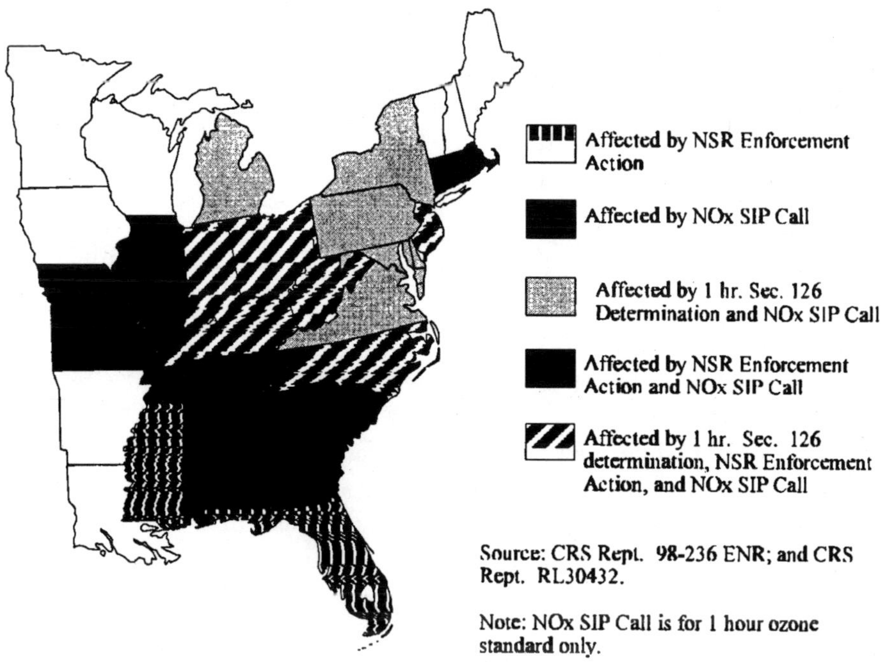

Figure 1.

The costs and benefits of these initiatives could be substantial, as indicated by Table 1. The NO_x SIP Call is the most wide ranging of the initiatives, with estimated costs of $1.7 billion annually and estimated quantifiable benefits of $1.1-$4.2 billion annually. Because EPA's methodology uses cost-effectiveness for determining emission budgets, the lion's share of the costs would be borne by the utility industry. The smaller scope of the Section 126 determinations reduces emissions abatement and benefits, but also costs. Of course, this scope could increase if additional petitions submitted to EPA result in more states being implicated as sources of transported ozone. Finally, the evolving scope of EPA's NSR action makes estimates of its costs and benefits difficult, if not impossible, at this time.

Table 1. Estimated Costs and Benefits of Initiatives

	Estimated Emission Reductions (2007)	**Estimated Costs (1990$)**	**Estimated Quantifiable Benefits (1990$)**	**Estimated Costs to Utilities (1990$)**
NO_x SIP Call	NO_x –1.14 million tons (ozone season)	$1.7 billion	$1.1-$4.2 billion	$1.4 billion
Section 126 Findings	NO_x – 0.66 million tons (ozone season)	$1.0 billion	$1.2 billion	$1.0 billion
NSR Action	NO_x – n/a SO_2 – n/a Particulates – n/a (annually)	n/a	n/a	n/a

n/a = not available

Source: CRS Report 98-236.

RECENT ACTIONS

Since January 2000 significant actions have occurred with all three of the initiatives. The status of these initiatives as of January 22, 2002, is summarized in Table 2. Perhaps the most significant action has been the decision of a 3-judge panel of the D.C. Circuit Court of Appeals to uphold EPA's Ozone Transport Rule with respect to the 1-hour ozone NAAQS (*Michigan v. EPA*, No. 98-1497 (D.C. Cir. March 3, 2000)), and to lift the stay on implementation. In upholding EPA's authority and methodology in developing the NO_x SIP Call, the court did make some modifications; in particular, that EPA's methodology did not support the inclusion of Wisconsin or all of Missouri and Georgia in the Rule (a decision reflected in Figure 1). In lifting the stay, the court ordered affected states to submit revised State Implementation Plans (SIPs) within 4 months of its June 22, 2000, order. In a subsequent ruling issued August 30, 2000, the court ordered EPA to move its original May 2003, compliance deadline to May 31, 2004. In March 2001, the Supreme Court denied a hearing to opponents of the SIP Call, effectively affirming the appeals court decision. In December 2000, EPA declared that 11 states and the District of Columbia failed to submit revised SIPs by the extended October 30th, 2000 deadline. By November 2002, all the affected states had submitted revised SIPs except Michigan, which has submitted a draft SIP revision.

None of these proceedings, however, affect the indefinite stay of EPA findings with respect to the 8-hour ozone standard. In February 2001, the Supreme Court ruled that although EPA has the authority to set a new 8-hour ozone standard, its interpretation of the relationship between the 1-hour standard's statutory implementation strategy and its new 8-hour standard implementation strategy was unreasonable and unlawful. The Court left it to EPA to "develop a reasonable interpretation" of the statutory provisions as they relate to the implementing the new 8-hour standard (*Whitman v. American Trucking Associations*, 531 U.S. 457 (2001) decided February 27, 2001).

With respect to the Section 126 petitions, EPA announced its 1-hour ozone findings on the 8 original petitions on January 18, 2000.[3] EPA granted four of the eight petitions for the 1-hour ozone standard: Connecticut, Massachusetts, New York, and Pennsylvania. Petitions from four other states were denied as these states no longer had areas that were not in attainment with the 1-hour standard. The rule specifies NO_x allocations for 392 facilities in 12 states and the District of Columbia, and implemented through a cap-and-trade program. The D.C. Circuit Court of Appeals upheld EPA's authority to issue the rule on May 15, 2001, but ordered EPA to reconsider factors used in setting emission limits (*Appalachian Power Co. v. EPA*). EPA responded to the court's order on August 3, 2001. On January 15, 2002, the EPA announced it would delay the compliance deadline for the Section 126 rule from May 1, 2003, to May 31, 2004, in line with the deadline for the NO_x SIP Call. EPA argues that a court order issued August 24, 2001, had already suspended the compliance deadline for powerplants, and it would be unfair to make other emission sources meet an earlier deadline.

Table 2. Status of Initiatives

<table>
<tr><th></th><th>Ozone Transport Rule</th><th>Section 126 Petitions</th><th>NSR Enforcement</th></tr>
<tr><td>1-hour Ozone Standard</td><td>Supreme Court denies hearing on SIP Call 3/5/01. Stay lifted by appeals court 6/22/00. EPA finds 11 states and D.C. failed to submit SIPs 11/19/00. Deadline for compliance moved to March 31, 2004.</td><td>EPA granted 4 of original petitions, 1/18/00. D.C. Circuit Court of Appeals upholds EPA authority, 5/15/01.

On 1/15/02, EPA announces it will delay compliance deadline until March 31, 2004.</td><td rowspan="2">In February 2000, EPA and Tampa Electric announced agreement on a consent decree that would settle EPA's NSR suit with that utility. In November and December 2000, EPA announced agreements in principle with Virginia Power and Cinergy.

In January 2002, EPA and PSEG announced the filing and settlement of a EPA/ State of New Jersey NSR suit.</td></tr>
<tr><td>8-hour Ozone Standard</td><td>Appeals court granted EPA motion to stay 8-hour findings. Supreme Court rules EPA has authority to promulgate 8-hour standard, but that its implementation strategy is unlawful. EPA left to develop a "reasonable" alternative.</td><td>EPA has indefinitely stayed finding on original petitions, 1/18/00.</td></tr>
</table>

In January 2000, EPA decided to indefinitely stay its original final determinations with respect to the 8-hour standard, given litigation regarding that standard. It also announced that

[3] Environmental Protection Agency, *Findings of Significant Contribution and Rulemaking on Section 126 Petitions for Purposes of Reducing Interstate Ozone Transport: Final Rule*, 65 *Federal Register* 2675-2767, January 18, 2000.

findings with respect to petitions by the District of Columbia, Delaware, Maryland, and New Jersey would be determined in the future. Since the filing of the NSR lawsuits in November 1999 (and subsequent lawsuits filed in 2000), several significant actions have occurred. First, in February 2000, EPA announced that it had come to an agreement with Tampa Electric Company on a consent decree that will settle the NSR lawsuit against that utility. The agreement will reduce NO_x emissions by over 85% (and SO_2 emissions by almost 80%) through a combination of fuel switching to natural gas, pollution control equipment optimization, and other techniques. The estimated $1 billion program is expected by Tampa Electric to have a "small" impact on its customers' bills.[4]

Second, on November 16, 2000, EPA and Virginia Power announced that an "agreement in principle" had been reach to settle EPA's NSR suit against Virginia Power. Over 12 years, Virginia will spend $1.2 billion to reduce NO_x and SO_2 emissions by about 70% through a combination of pollution control equipment and fuel switching. This announcement was followed on December 21, 2000 by a similar agreement in principle between EPA and Cinergy involving a $1.4 billion investment in control technology.

Third, on January 24, 2002, EPA and the State of New Jersey announced the filing and settlement of an NSR suit against PSEG Fossil LLC. PSEG agreed to reduce its SO_2 emissions by 90% and its NO_x emissions by 83% from 2000 levels by 2012 at an estimated cost of $337 million. In addition, PSEG agreed to reduce CO_2 emissions by 15% from 1990 levels.

Litigation on these cases has slowed as participants assess the impact of the Bush Administration's June 2002 recommendations to revise the New Source Review process. Of particular interest is an EPA recommendation that a new rulemaking be commenced on the definition of "routine maintenance," a key point of contention in the above lawsuits. As of October 2002, no formal rulemaking has been proposed by EPA as the drafting process has not been completed.[5]

Congressional Actions

The continuing difficulties in the Northeast both in meeting the ozone NAAQS and in reducing acid precipitation have focused attention on emissions from fossil fuel-fired utilities, particularly of NO_x – and on the potential costs of reducing those emissions. Concerned about

---

[4] Tampa Electric, "Tampa Electric Reaches Agreement with EPA, Department of Justice on Environmental Issues," [http://www.tampaelectric.com/energy/ENEPAagrmnt.html].

[5] The seven recommendations fall into two categories: (1) four recommendations that would complete a rulemaking process begun under the Clinton Administration in 1996 and would be issued as a direct final rule; and (2) three recommendations that would be put out as a proposed rulemaking and subject to normal administrative procedures. Revisions to the definition of routine maintenance would fall under the second category.

the piecemeal nature of these initiatives, some in Congress have been working on comprehensive, multi-pollutant alternative strategies to reduce emissions. In June 2002, the Senate Environment and Public Works Committee reported out S. 556 – a comprehensive, multi-pollutant bill that would incorporate market-oriented mechanisms to control NO_x, SO2, and CO2, and tonnage limitations on Hg.[6] No floor action has been scheduled.

[6] For a review of proposed legislation, see Larry Parker, *Air Quality: Multi-pollutant Legislation*, CRS Report RL31326, October 22, 2002.

Chapter 3

CLEAN AIR ACT ISSUES[*]

James E. McCarthy

INTRODUCTION

The most prominent air quality issue in recent months has been the controversy over EPA's proposed changes to the New Source Review (NSR) requirements, which impose emission controls on modifications of power plants and other major facilities. In its consideration of the omnibus FY2003 appropriation bill (H.J.Res. 2) on January 22, the Senate narrowly defeated an amendment that would have delayed implementation of changes to the requirements pending a study by the National Academy of Sciences (NAS). The Senate did approve a separate amendment directing NAS to conduct such a study, but not delaying implementation of the standards. The President signed the bill, with the latter amendment, February 20 (P.L. 108-7).

In addition to changing NSR, the Administration has asked Congress to modify Clean Air Act requirements for power plants by enacting "Clear Skies" or "multi-pollutant" legislation. In the 107th Congress, the Senate Environment and Public Works Committee narrowly approved multi-pollutant legislation (Senator Jeffords' S. 556) June 27, 2002; but the Administration and much of the electric power industry opposed the bill, and it did not reach the Senate floor. "Clear Skies" legislation (S. 485 / H.R. 999) was reintroduced in the 108th Congress, February 27. Senator Jeffords and Senator Carper have also introduced bills (S. 366 and S. 843 respectively). The Clean Air subcommittee of Senate Environment and Public Works held hearings on Clear Skies April 8 and May 8 and another hearing is expected in June.

Another holdover issue from previous Congresses concerns regulation of the gasoline additive MTBE. MTBE is used to meet Clean Air Act requirements that gasoline sold in the nation's worst ozone nonattainment areas contain at least 2% oxygen, to improve combustion. The additive has been implicated in numerous incidents of ground water contamination, however, and 17 states have taken steps to ban or regulate its use. The most significant of

---

[*] Excerpted from CRS Report IB10107.

these bans (in California and New York) take effect at the end of 2003, leading many to suggest that Congress revisit the issue before then to modify the oxygenate requirement and set more uniform national requirements regarding MTBE and its potential replacements (principally ethanol).

The comprehensive energy bill that passed the House April 11 (H.R. 6) addresses some of these issues, eliminating the oxygen requirement, providing funds for the cleanup of MTBE in ground water and for conversion of MTBE production facilities, and requiring the use of renewable fuels such as ethanol in gasoline. It does not ban the use of MTBE, however. On April 9, the Senate Environment and Public Works Committee ordered similar legislation (S. 791) to be reported. S. 791 *would* ban MTBE use in motor fuels 4 years after the date of enactment, except in states that specifically authorize its use. Provisions similar to S. 791 are expected to offered as an amendment to the Senate energy bill, S. 14.

Other clean air issues that might be considered in the 108th Congress are the conformity of metropolitan area transportation plans with the Clean Air Act, and whether to modify the Act's requirements for areas that have not met deadlines for attainment of the ozone air quality standard.

Most Recent Developments

On April 11, the House passed H.R. 6, a comprehensive energy bill. Among its provisions were amendments to the Clean Air Act's reformulated gasoline (RFG) program, eliminating the requirement that RFG contain 2% oxygen and establishing a new requirement that an increasing percentage of gasoline contain renewable fuels such as ethanol. Two days earlier, the Senate Environment and Public Works Committee ordered similar provisions reported in S. 791. The Senate bill would also provide a ban on the use of MTBE in motor fuels, except in states that specifically authorize its use.

Bills to regulate emissions of multiple pollutants from electric power plants, including the Administration's Clear Skies bill and several competing bills, have been introduced in both the House and Senate. The Clean Air subcommittee of Senate Environment and Public Works held hearings on Clear Skies April 8 and May 8, and plans another in June.

On January 22, the Senate narrowly defeated an amendment to the FY2003 Omnibus Appropriations bill (H.J.Res. 2) that would have delayed implementation of changes to the Clean Air Act's New Source Review requirements pending completion of a study by the National Academy of Sciences. Environmental Protection Agency (EPA) recommendations regarding the NSR program were released November 22, 2002, and appeared in the Federal Register December 31. The new NSR rules (some proposed and others promulgated by EPA) will make it easier for companies to modify their facilities without installing new pollution controls.

Background and Analysis

Despite steady improvements in air quality in many of the United States' most polluted cities, the goal of clean air continues to elude the nation: 107 areas with a combined population of 97.8 million were classified as "nonattainment" for one or more of the National

Ambient Air Quality Standards (NAAQS) as of December 2002. Two pollutants account for the vast majority of nonattainment areas: ozone — 36 areas with 85.5 million people — and particulate matter (PM) — 61 areas with 24.9 million people. Thirty-nine areas with 18.4 million people have failed to achieve standards for carbon monoxide, sulfur dioxide, or lead.

The standards for these pollutants are health-based: the statute requires that EPA set them at levels necessary to protect the public health with an adequate margin of safety, based on a review of the scientific literature. From time to time (every 5 years according to the statute, but less frequently in reality), the Agency reviews the latest scientific studies and either reaffirms or modifies the standards. The most recent changes (a strengthening of the ozone and PM standards) were promulgated in 1997. Due to legal challenges and other delays, the new standards have not yet been implemented. When they are implemented (now expected in 2004), they are likely to double the number of areas in nonattainment.

National Ambient Air Quality Standards drive many of the Clean Air Act's programs. The need to attain them sets in motion State Implementation Plans that establish detailed requirements for sources of air pollution, including: the imposition of Reasonably Available Control Technologies on stationary sources of pollution; the requirement that new sources of pollution in nonattainment areas "offset" their emissions by reductions in pollution from other sources; the operation of inspection and maintenance programs for auto emission controls; the requirement to use cleaner burning reformulated gasoline as a means of reducing emissions; and the necessity of demonstrating that new highway and transit projects "conform" to the State Implementation Plan for the area in which they will be constructed.

Other provisions of the Act are separate from the State Implementation Plans, and are for the most part national in scope. These include emission standards for cars, trucks, and other mobile sources of pollution; standards for new major stationary sources of pollution; emission standards for sources of hazardous air pollutants; standards for prevention of significant deterioration in areas where air quality is better than the NAAQS; acid rain and regional haze programs; and stratospheric ozone provisions.

Issues in the 108th Congress

Several of the clean air issues facing the 108th Congress are holdovers that were discussed at length, but not resolved, in the 107th. Changes to the Act's reformulated gasoline program, for example, including a ban on use of the gasoline additive MTBE, reached a conference committee as part of the comprehensive energy bill (H.R. 4); and legislation was reported on the regulation of emissions from electric power plants (S. 556). Ultimately, neither bill was enacted, leaving these issues for consideration in the 108th.

In the remainder of this Issue Brief, we look in more detail at five prominent air issues that might be of interest in the 108th Congress: New Source Review, multi-pollutant legislation, MTBE, transportation conformity, and deadlines for achieving the ozone air quality standard.

New Source Review (NSR)

The most prominent air quality issue in recent months has been whether to modify the Clean Air Act's New Source Review requirements. EPA proposed and promulgated several changes to these rules on December 31, 2002, the net effect of which will be to allow modification of some existing major sources of air pollution without subjecting them to new emission standards.

The controversy over the NSR process stems from EPA's application of New Source Performance Standards to existing stationary sources of air pollution that have been modified. In Section 111, the Clean Air Act states that new sources (subject to NSR) include modifications of existing sources as well as plants that are totally new. Industry has generally avoided the NSR process, however, by claiming that changes to existing sources were "routine maintenance" rather than modifications. In the 1990s, EPA began reviewing records of electric utilities, petroleum refineries, and other industries to determine whether the changes were routine. As a result of these reviews, since late 1999, EPA and the Department of Justice have filed suit against 15 electric utilities, claiming that they made major modifications to 58 plants in 15 states, extending their lives and increasing their electric generating capacity without undergoing required New Source Reviews and without installing best available pollution controls. With two exceptions, these suits were filed by the Clinton Administration.

Four of the 15 utilities charged with NSR violations (Tampa Electric, PSEG of New Jersey, Dominion Resources/Virginia Electric Power, and Wisconsin Electric Power) have settled with EPA, agreeing to spend more than $3.1 billion over the next decade on pollution controls or fuel switching in order to reduce emissions at their affected units. One other utility (Cinergy) reached agreement in principle more than two-and-a-half years ago to spend more than $1 billion to resolve NSR violations, but final settlement negotiations have not been concluded. A sixth utility, the Tennessee Valley Authority, has announced plans to spend $1.5 billion to reduce emissions at four of its plants, although not as part of a settlement agreement. Between July 25, 2000 and December 20, 2001, the Agency also reached agreement with nine petroleum refiners representing more than 30% of industry capacity. The refiners agreed to settle potential charges of NSR violations by paying fines and installing equipment to eliminate 153,000 tons of pollution.

Most of the utilities have not settled with EPA. They and other critics of the Agency's enforcement actions claim that EPA reinvented the rules. A strict interpretation of what constitutes routine maintenance, they contend, will prevent them from making changes that were previously allowed, without a commitment of time and money for permit reviews and the installation of expensive pollution control equipment. This provides disincentives for power producers, refiners, and others to expand output at existing facilities, they maintain.

NSR rules have been under review by the Administration since the May 2001 release of the President's energy plan. In the plan, EPA was charged with undertaking a review of NSR, with the EPA Administrator to report to the President regarding the impact of NSR regulations on investment in new utility and refinery generation capacity, energy efficiency, and environmental protection. The review was concluded in June 2002, and regulatory changes were promulgated and proposed December 31, 2002.

EPA promulgated four sets of changes to NSR. First, it will allow facilities to use Plantwide Applicability Limits, rather than emissions from the individual units being replaced, to determine whether emissions will increase from a plant modification (this is expected to make it easier to modify facilities without triggering NSR). Second, certain environmentally beneficial pollution control and prevention projects will be allowed to proceed without NSR permits, upon submission of a notice to the permitting authority. Third, plants that install state-of-the-art pollution controls (referred to as "clean units") will be allowed to modify their facilities during the ensuing 10 years without undergoing further review, provided they meet emission limits specified in their permit. And fourth, the methodology used to calculate whether emissions will increase (triggering NSR) will be changed — for example, facilities other than power plants will be able to compare projected emissions after a modification to the highest emission levels reached during any 24-month period during the previous 10 years.

In addition to the four promulgated changes, the Agency also proposed new regulations defining what constitutes routine maintenance, which is exempt from review. The proposal would exempt from NSR modifications that cost less than threshold amounts.

The proposed and promulgated changes have been characterized by the Administration as a streamlining or improvement of the program, and by environmental groups and a number of states as a significant weakening. On the day they were promulgated, nine Northeastern states filed suit to overturn the changes; thus, enforcement of NSR is likely to be blocked for the immediate future. In the meantime, the prospect of an NSR rollback, critics argue, has caused utilities to withdraw from settlement negotiations over the pending lawsuits, delaying emission reductions that could have been achieved in the near future.

On January 22, the Senate narrowly defeated an amendment offered by Senator Edwards (S.Amdt. 67 to H.J.Res. 2) that would have delayed implementation of changes to the NSR requirements for 6 months pending a study by the National Academy of Sciences. The Senate did approve a separate amendment offered by Senator Inhofe (S.Amdt. 86) directing NAS to conduct such a study, but not delaying implementation of the standards. The amendment was enacted as Section 356 of the Omnibus Appropriations bill (P.L. 108-7).

Perhaps further complicating the issue, on April 21, 2003, the National Academy of Public Administration released a report commissioned by Congress that made sweeping recommendations to modify NSR. The report concluded that the NSR permitting process works as Congress intended for new industrial facilities, but has not been effective in reducing air pollution when changes at existing sources are likely to increase emissions. "Instead — contrary to Congressional intent — many large, highly polluting facilities have continued to operate and have expanded their production (and pollution) over the past 25 years without upgrading to cleaner technologies," the report states. The study panel recommended that Congress end the "grandfathering" of major air emission sources, by requiring all major sources that have not obtained an NSR permit since 1977 to install Best Available Control Technology or Lowest Achievable Emissions Rate control equipment. In the interim, the NAPA panel concluded, EPA and the Department of Justice should continue to enforce NSR vigorously, especially for changes at existing facilities.

Clear Skies / Multi-Pollutant Legislation

In addition to its proposed and promulgated regulatory changes in NSR, the Administration has asked Congress to modify Clean Air Act requirements for power plants by enacting "Clear Skies" or "multi-pollutant" legislation. A number of multi-pollutant bills have been introduced.

Depending on the bill's author, such legislation comes in 3- or 4-pollutant versions. The 3-pollutant bills would set standards for sulfur dioxide, nitrogen oxides, and mercury. The 4-pollutant bills add carbon dioxide to the mix.

Such legislation, it is argued — whether in 3- or 4-pollutant form — would both reduce emissions and encourage investment in new plants by providing certainty regarding future regulatory requirements. In some proposed bills, the new requirements would replace numerous existing regulatory programs, including NSR, New Source Performance Standards, Prevention of Significant Deterioration, Lowest Achievable Emission Rate standards, Best Available Retrofit Technology, and regulations under development to control mercury emissions from electric utilities.

The number of these current and prospective regulations on power plant emissions has suggested to many in industry, environmental groups, Congress, and the Administration that the time is ripe for such comprehensive legislation. The key questions are how stringent the controls will be, and whether carbon dioxide (CO_2) will be among the emissions subject to controls.

Regarding the stringency issue, bills introduced in the 108th Congress would require reduction of NO_x emissions to 1.5 or 1.7 million tons per year (a 70% - 80% reduction from 1998 levels) and reduction of sulfur dioxide emissions to 2.23 - 3.0 million tons per year (also a reduction of 70% - 80% versus 1998). Regarding mercury, the bills would either require EPA to determine the level of reductions, or require reductions of 70% - 90% from current levels of emissions (from 48 to 5, 10, or 15 tons annually, depending on the bill). In the most stringent of the bills (Sen. Jeffords' S. 366 and Rep. Waxman's H.R. 2042), these reductions would take place by 2008 or 2009. Three of the bills (Sen. Jeffords', Rep. Waxman's, and Sen. Carper's S. 843) would also set caps on CO_2 emissions, at the level emitted in 1990 or 2000.

The Administration's "Clear Skies" bill (H.R. 999 / S. 485) envisions less stringent standards than those in the other bills, phased in over a longer period of time. For NO_x, the Administration would reduce emissions to 1.7 million tons per year by 2018, with an intermediate limit of 2.1 million tons in 2008. For sulfur dioxide, the limit would be 3.0 million tons annually in 2018, with an intermediate limit of 4.5 million tons in 2010. For mercury, the limit would be 26 tons per year in 2010, declining to 15 tons in 2018. "Clear Skies" and most of the other bills envision a system like that used in the acid rain program, where national or regional caps on emissions are implemented through a system of tradeable allowances.

The Administration opposes controls on CO_2, viewing them as a step towards implementing the Kyoto Protocol to the United Nations Framework Convention on Climate Change, which it opposes. The absence of CO_2 from the mix leads to different strategies for achieving compliance, preserving more of a market for coal, and lessening the degree to which power producers might switch to natural gas or renewable fuels as a compliance strategy.

Four hearings on multi-pollutant legislation were held by the Senate Environment and Public Works Committee in the 107th Congress, and the Committee narrowly approved Senator Jeffords' 4-pollutant bill, with amendments, June 27, 2002 (S. 556, S.Rept. 107-347). Opposed by the Administration and by the electric utility and coal industries, the bill died without reaching the Senate floor. Senator Jeffords has reintroduced this bill in the 108th Congress as S. 366.

In addition to the Jeffords bill and the other bills discussed above, Senator Carper introduced legislation late in the 107th Congress that was described as an effort to reach middle ground. Reintroduced in the 108th Congress with minor changes as S. 843, the Carper bill would regulate CO_2 as well as the other three pollutants. Its deadlines and the required reductions in emissions fall somewhere between the Jeffords bill and the Administration's Clear Skies proposal.

Prospects for Clear Skies and other multipollutant bills are complicated. The House, with its larger Republican majority and more formal rules, could presumably pass Clear Skies if the leadership decided to make it a priority. In the Senate, however, consensus has yet to emerge in the Environment and Public Works Committee, to which the bill and other multipollutant bills have been referred. At least at present, Clear Skies does not enjoy the support of a majority of the Committee's members. Faced with this obstacle, some have suggested that the Senate leadership take the bill directly to the Senate floor, bypassing the committee; but it might face determined opposition there, as well. For now, the Environment Committee is proceeding with hearings and the Administration continues to say the bill is a high priority, but the prospects for action remain unclear.

MTBE

Another holdover issue from previous Congresses concerns regulation of the gasoline additive MTBE (methyl tertiary butyl ether). This issue appears to be on a faster track, with the House having passed legislation to address it in April, and the Senate likely to take the issue up (as part of the energy bill debate) in June.

MTBE is used to meet Clean Air Act requirements that reformulated gasoline (RFG), sold in the nation's worst ozone nonattainment areas, contain at least 2% oxygen, to improve combustion. Under the RFG program, areas with "severe" or "extreme" ozone pollution (82 counties with a combined population of 55 million) must use reformulated gas; areas with less severe ozone pollution may opt into the program as well, and many have. In all, portions of 17 states and the District of Columbia use RFG, and about 30% of the gasoline sold in the United States is RFG.

The law requires that RFG contain at least 2% oxygen by weight. Refiners can meet this requirement by adding a number of ethers or alcohols, any of which contains oxygen and other elements. By far the most commonly used oxygenate is MTBE. In 1999, 87% of RFG contained MTBE, a number since reduced to about 70%. MTBE has also been used since the late 1970s in non-reformulated gasoline, as an octane enhancer, at lower concentrations. As a result, gasoline with MTBE has been used virtually everywhere in the United States, whether or not an area has been subject to RFG requirements.

MTBE leaks, generally from underground gasoline storage tanks, have been implicated in numerous incidents of ground water contamination. The substance creates taste and odor problems in water at very low concentrations, and some animal studies indicate it may pose a potential cancer risk to humans. For these reasons, 17 states have taken steps to ban or regulate its use. The most significant of the bans (in California and New York) take effect at the end of 2003, leading many to suggest that Congress revisit the issue before then to modify the oxygenate requirement and set more uniform national requirements regarding MTBE and its potential replacements (principally ethanol).

Support for eliminating the oxygen requirement on a nationwide basis is widespread among environmental groups, the petroleum industry, and states. In general, these groups have concluded that gasoline can meet the same low emission performance standards as RFG without the use of oxygenates. But a potential obstacle to enacting legislation to remove the oxygen requirement lies among agricultural interests. Nearly 10% of the nation's corn crop is used to produce the competing oxygenate, ethanol. If MTBE use is reduced or phased out, but the oxygen requirement remains in effect, ethanol use would soar, increasing demand for corn. (In fact, ethanol use is already growing as MTBE begins to be phased out.) Conversely, if the oxygen requirement is waived by EPA or legislation, not only would MTBE use decline, but so, likely, would demand for ethanol. Thus, Members of Congress and Governors from corn-growing states have taken a keen interest in MTBE legislation. Unless their interests are addressed, they could pose a potent obstacle to its passage.

Relying heavily on legislation that reached a conference committee in the 107th Congress, the 108th Congress has moved quickly to address the MTBE and ethanol issues. On April 11, the House passed H.R. 6, a comprehensive energy bill. Title VII of the bill would amend the Clean Air Act to eliminate the requirement that RFG contain 2% oxygen and establish a new requirement that an increasing amount of gasoline contain renewable fuels such as ethanol. The bill would require more than a doubling of ethanol use by 2015. The bill would also authorize $850 million for MTBE cleanup, authorize $750 million to assist the conversion of merchant MTBE production facilities to the production of other fuel additives, preserve the reductions in emissions of toxic substances achieved by the RFG program, and provide a "safe harbor" from liability lawsuits for producers of MTBE. The bill would not ban the use of MTBE, however. Two days earlier, the Senate Environment and Public Works Committee ordered similar, but not identical provisions reported in S. 791. Among the differences, the Senate bill *would* ban the use of MTBE in motor fuels 4 years after the date of enactment, except in states that specifically authorize its use. It also would not provide a safe harbor for MTBE producers. Provisions similar to S. 791 are expected to be offered on the Senate floor during debate on S. 14, the Senate's comprehensive energy bill.

As the deadlines for state phaseout of MTBE move closer, investment decisions involving hundreds of millions of dollars hang on the regulatory framework of the post-MTBE gasoline market. Thus, pressure for congressional action on this issue is likely to remain high. Whether this pressure will produce enacted legislation is less clear.

Conformity of Transportation Plans and SIPs

A fourth clean air issue that might be considered in the 108th Congress is the conformity of metropolitan area transportation plans with the Clean Air Act. Under the Act, areas that have not attained any of the six National Ambient Air Quality Standards must develop State Implementation Plans (SIPs) demonstrating how they will reach attainment. As of December 2002, 107 areas with a combined population of 97.8 million people were subject to the SIP requirements. Section 176 of the Clean Air Act prohibits federal agencies from funding projects in these areas unless they "conform" to the SIPs. Specifically, projects must not "cause or contribute to any new violation of any standard," "increase the frequency or severity of any existing violation," or "delay timely attainment of any standard." Because new highways generally lead to an increase in vehicle miles traveled and related emissions, both the statute and regulations require that an area's Transportation Improvement Program (TIP), which identifies major highway and transit projects an area will undertake, obtain a new demonstration of conformity no less frequently than every 3 years. Highway and transit projects cannot receive federal funds unless they are part of a conforming TIP.

The impact of conformity requirements is expected to grow in the next several years for several reasons. The growth of emissions from SUVs and other light trucks and greater than expected increases in vehicle miles traveled have both made it more difficult to demonstrate conformity; recent court decisions have tightened the conformity rules; and the scheduled implementation of more stringent air quality standards in 2004 will mean that additional areas are subject to conformity. Thus, numerous metropolitan areas will face a cutoff of highway and transit funds unless they impose sharp reductions in vehicle, industrial, or other emissions. In a recent survey, the General Accounting Office found that, over the past 6 years, only 5 metropolitan areas have had to change transportation plans in order to resolve a conformity lapse; but about one-third of local transportation planners surveyed expect to have difficulty demonstrating conformity in the future. (See U.S. GAO, *Environmental Protection: Federal Planning Requirements for Transportation and Air Quality Protection Could Potentially Be More Efficient and Better Linked*, April 2003, for additional detail.)

Of particular concern to areas facing a potential conformity lapse may be the fact that the Clean Air Act provides no authority for waivers or grace periods during a lapse. Only a limited set of exempt projects (mostly safety-related or replacement and repair of existing transit facilities) can be funded in lapsed areas: the rules do not even allow funding of new projects that might reduce emissions, such as new transit lines. These limitations are among the issues that may be raised by those seeking to amend the conformity provisions. In addition, many have raised concerns about a mismatch between the SIP and TIP planning cycles, and have called for less frequent, but better coordinated demonstrations of conformity. In its recent report, the General Accounting Office recommended that "relevant federal agencies (1) consider extending the 3-year time frame between required transportation plan updates and asking the Congress to amend the Clean Air Act to change the conformity rules to match, and (2) assess the advantages and disadvantages of statutorily requiring that the emissions budgets in air quality plans be regularly updated with new travel data and emissions models." At least the first of these recommendations appears to be generally supported by transportation planners and highway builders, but opposed by environmental groups and air quality planning officials.

In the 108th Congress, H.R. 673 would repeal the existing conformity regulations and require EPA to promulgate revised criteria and procedures for conformity within one year of enactment. Conformity provisions have also been introduced in the Administration's highway and transit legislation (H.R. 2088 / S. 1072); Section 6001 of the bill would require conformity demonstrations every 5 years instead of every three, and would shorten the planning horizon over which conformity must be demonstrated to 10 years in most cases, instead of the current 20 years.

Conformity issues have also been raised by the Department of Defense in the Pentagon's Readiness and Range Preservation Initiative, discussed as part of the fiscal year 2004 defense authorization bill (H.R. 1588, S. 1050) currently moving through Congress. (H.R. 1588 passed the House May 22; S. 1050 passed the Senate the same day.) As part of an effort to ease environmental restrictions on the use of active military ranges, DOD proposed that the Department be allowed to conduct military readiness activities that might otherwise be prohibited under the conformity requirement. The proposal would have required that the Department estimate the quantity of emissions caused by the readiness activities, notify the state air quality agency before engaging in such activities, and ensure conformity within 3 years of the date new activities began. It also would have provided that nonattainment areas that would be eligible for reclassification as "attainment" but for the emissions caused by readiness activities be allowed to subtract such emissions from the areas' emission inventories. While these and other elements of the readiness and range initiative were much discussed, neither the House nor the Senate bill contains the conformity provisions. In hearings, many Members indicated that DOD had not provided sufficient justification and requested that the Department provide quantitative data in support of its request. As a result, as passed by the House, H.R. 1588 provides for the Secretary of Defense to conduct a study on the impact, if any, of compliance by the Department of Defense with State Implementation Plans for air quality under the Clean Air Act.

Deadlines for Achieving the Ozone and PM Air Quality Standards

A fifth set of issues that were discussed, in part, as a possible amendment to the Omnibus Appropriations bill in early January concerns the deadlines for achievement of the ozone air quality standard. Under the 1990 Clean Air Act Amendments, ozone nonattainment areas were classified in one of five categories (Marginal, Moderate, Serious, Severe, or Extreme) depending on the concentration of ozone recorded by air quality monitoring equipment in the 3 years preceding passage of the 1990 amendments. Areas with higher concentrations of the pollutant were required to implement more stringent controls on emissions; they were also given more time to reach attainment. Failure to reach attainment by the specified deadline was to result in reclassification of an area to the next highest category and the imposition of more stringent controls. Areas classified as Serious, for example, were required to reach attainment by 1999. If they did not do so, the law requires that they be reclassified as Severe, with a new deadline of 2005, and more stringent emission controls, including the imposition of controls on smaller sources of air pollution.

For a variety of reasons, EPA has generally not reclassified areas when they failed to reach attainment by the statutory deadlines. The Agency's website currently lists 21 Marginal areas, 9 Moderate areas, and 14 Serious areas, most of which should be categorized as Severe had the Agency adhered to the statutory requirements. In many cases, the Agency granted additional time to reach attainment on the grounds that a major cause of an area's continued nonattainment was pollution generated outside the area and transported into it by prevailing winds. The Agency has been sued over its failure to reclassify three areas (Washington, D.C., St. Louis, and Beaumont-Port Arthur, Texas) and has lost all three suits. As a result, numerous areas are expected to be subject to reclassification, with more stringent pollution control requirements, in the near future.

In early January, reports surfaced that Texas environmental officials had written House Members asking for an amendment that would allow EPA to grant nonattainment areas additional time to achieve the ozone standard. No amendment was introduced, however, and the issue appears to have been set aside for now.

Another deadline issue concerns the implementation of new standards for ozone and fine particles that EPA promulgated in 1997. Due to legal challenges and other delays, the new standards have not yet been implemented, but when they are implemented (now expected in 2004), they are likely to double the number of areas in nonattainment. In response to an initiative from the State of Texas, in 2002, EPA approved a protocol under which areas can avoid designation as nonattainment for ozone until December 31, 2007, if they voluntarily commit to enforceable early action compacts with their state and EPA. The protocol sets out a number of milestones that areas must meet to qualify. Thirty-five areas met the first of these requirements by submitting signed compacts to EPA by December 31, 2002.

The Administration has proposed an additional modification of the requirements for areas not meeting the new ozone and fine particle standards in its Clear Skies bill (H.R. 999 / S. 485). In Section 3, Clear Skies would allow EPA to avoid designating 8-hour ozone and

$PM_{2.5}$ areas as nonattainment until 2016, provided that the area demonstrates that it will attain the standards by December 31, 2015. Areas fitting into this new "transitional" category could avoid additional regulatory controls, including the requirement to demonstrate conformity, if they could demonstrate that attainment will be achieved through the imposition of federal controls on utilities, diesel engines, automobiles, and other sources.

LEGISLATION

H.R. 6 (Tauzin)

Energy Policy Act of 2003. Title VII amends the Clean Air Act to remove the oxygen content requirement for RFG, to increase production and use of renewable fuels such as ethanol, to provide a "safe harbor" from lawsuits for producers of renewable fuels and MTBE, to provide assistance for conversion of merchant MTBE production facilities, and to prevent backsliding on emissions of toxic air pollutants from RFG. Also amends the Solid Waste Disposal Act to authorize funding for cleanup of MTBE. Incorporates provisions of H.R. 1644 (Barton), reported April 8, 2003 (H.Rept. 108-65, Part 1 — see Title IX). H.R. 6 introduced April 7, 2003; referred to the Committee on Energy and Commerce, and in addition to the Committees on Science, Ways and Means, Resources, Education and the Workforce, Transportation and Infrastructure, Financial Services, and Agriculture. Passed the House, 247-175, April 11, 2003.

H.R. 185 (Serrano)

Amends the Internal Revenue Code of 1986 to provide a business credit relating to the use of clean-fuel vehicles by businesses within areas designated as nonattainment areas under the Clean Air Act. Introduced January 7, 2003; referred to Committee on Ways and Means

H.R. 203 (Sweeney)

Reduces acid deposition under the Clean Air Act, and for other purposes. Introduced January 7, 2003; referred to Committee on Energy and Commerce.

H.R. 244 (Issa)

Amends the Clean Air Act to permit the exclusive application of California State regulations regarding reformulated gas in federal RFG areas within the State. Introduced January 7, 2003; referred to Committee on Energy and Commerce.

H.R. 427 (Sensenbrenner)

Fuel Price Stability Act of 2003. Amends the Clean Air Act to allow the Governors of Illinois, Indiana, and Wisconsin to permit the sale of conventional gasoline in reformulated gasoline areas if the Governor finds that reduced availability of RFG has resulted in, or is likely to result in, a significant price increase in that area. Introduced January 28, 2003; referred to Committee on Energy and Commerce.

H.R. 673 (K. Brady)

Safe Highways and Roads Act of 2003. Repeals the existing transportation conformity regulations, replacing them with those in effect prior to a March 1999 court decision, and requires EPA to promulgate revised criteria and procedures for conformity within one year of enactment. Introduced February 11, 2003; referred to Committee on Energy and Commerce.

H.R. 837 (C. Peterson)

Fuels Security Act of 2003. Amends the Clean Air Act to ban MTBE from the U.S. fuel supply not later than 4 years after the date of enactment, to eliminate the oxygen content requirement for reformulated gasoline while maintaining reductions in emissions of toxic air pollutants, to increase production and use of renewable fuels such as ethanol to 5 billion gallons per year by 2012, to provide a "safe harbor" from liability resulting from the use of renewable fuels, to require federal agencies to purchase gasoline containing at least 10% ethanol and diesel fuel containing biodiesel provided they are available at a generally competitive price, to authorize $400 million from the Leaking Underground Storage Tank Fund for remediation of MTBE contamination, and to authorize $750 million in grants for conversion of merchant MTBE production facilities. Introduced February 13, 2003; referred to Committee on Energy and Commerce.

H.R. 999 (Barton, by request)

Clear Skies Act of 2003. The Administration's multi-pollutant legislation for electric utility emissions of sulfur dioxide, nitrogen oxides, and mercury. Introduced February 27, 2003; referred to Committee on Energy and Commerce.

H.R. 1020 (P. Ryan)

Amends the Clean Air Act requirements relating to gasoline to prevent future supply shortages and price spikes in the gasoline market. Introduced February 27, 2003; referred to Committee on Energy and Commerce.

H.R. 1891 (Paul)

Amends the Clean Air Act to prohibit liability for the effects of emissions resulting from or caused by an act of nature including: volcanic eruptions and dust storms; accident; war; terrorism; or fires that occur beyond its jurisdiction related to land clearing, agriculture and ecological restoration and management. Introduced April 30, 2003; referred to Committee on Energy and Commerce and Committee on the Judiciary.

H.R. 2042 (Waxman)

Clean Smokestacks Act of 2003. Amends the Clean Air Act to reduce emissions of sulfur dioxide, nitrogen oxides, mercury, and carbon dioxide from electric powerplants. Introduced May 8, 2003; referred to Committee on Energy and Commerce.

H.R. 2136 (P. King)

Amends the Clean Air Act to prohibit the use of MTBE as a gasoline additive and to repeal the oxygenate requirement for reformulated gasoline, and to provide funding for the clean up of underground storage tanks. Introduced May 15, 2003; referred to Committee on Energy and Commerce.

H.R. 2253 (Pombo)

Amends the Clean Air Act to require EPA to ban the use of MTBE in gasoline as soon as practicable and to prohibit any gasoline additive unless it has been determined (through scientific testing and peer review) not to have any adverse effects on the public. Introduced May 22, 2003; referred to Committee on Energy and Commerce.

S. 366 (Jeffords)

Clean Power Act of 2003. Amends the Clean Air Act to reduce emissions of sulfur dioxide, nitrogen oxides, mercury, and carbon dioxide from electric powerplants. Introduced February 12, 2003; referred to Committee on Environment and Public Works.

S. 385 (Daschle)

Fuels Security Act of 2003. Amends the Clean Air Act to ban MTBE from the U.S. fuel supply not later than 4 years after the date of enactment, to eliminate the oxygen content requirement for reformulated gasoline while maintaining reductions in emissions of toxic air pollutants, to increase production and use of renewable fuels such as ethanol to 5 billion gallons per year by 2012, to provide a "safe harbor" from liability resulting from the use of renewable fuels, to require federal agencies to purchase gasoline containing at least 10% ethanol and diesel fuel containing biodiesel provided they are available at a generally competitive price, to authorize $400 million from the Leaking Underground Storage Tank Fund for remediation of MTBE contamination, and to authorize $750 million in grants for conversion of merchant MTBE production facilities. Introduced February 13, 2003; referred to Committee on Environment and Public Works.

S. 484 (Leahy)

Amends the Clean Air Act to establish requirements concerning the operation of fossil fuel-fired electric utility steam generating units, commercial and industrial boiler units, solid waste incineration units, medical waste incinerators, hazardous waste combustors, chlor-alkali plants, and Portland cement plants to reduce emissions of mercury to the environment. Introduced February 27, 2003; referred to Committee on Environment and Public Works.

S. 485 (Inhofe, by request)

Clear Skies Act of 2003. The Administration's multi-pollutant legislation for electric utility emissions of sulfur dioxide, nitrogen oxides, and mercury. Introduced February 27, 2003; referred to Committee on Environment and Public Works.

S. 791 (Inhofe)

Reliable Fuels Act. Amends the Clean Air Act to remove the oxygen content requirement for RFG, to eliminate MTBE from the U.S. fuel supply except in states that specifically authorize its use, to increase production and use of renewable fuels such as ethanol, to provide a "safe harbor" from lawsuits for producers of renewable fuels, and to prevent backsliding on emissions of toxic air pollutants from RFG. Also amends the Solid Waste Disposal Act to authorize funding for cleanup of MTBE. Introduced April 3, 2003; referred to Committee on Environment and Public Works. Ordered reported, with amendments, April 9, 2003.

S. 843 (Carper)

Amends the Clean Air Act to reduce emissions of sulfur dioxide, nitrogen oxides, mercury, and carbon dioxide from electric powerplants. Introduced April 9, 2003; referred to Committee on Environment and Public Works.

S.Amdt. 67 (Edwards)

Requires a study by the National Academy of Sciences of the effects of the final rule relating to New Source Review promulgated December 31, 2002, to determine whether it would result in any increase in air pollution or any adverse effect on human health. Delays implementation of EPA's changes to the NSR program for 6 months to allow completion of the study. Amendment was not agreed to, by a vote of 46 - 50. Record Vote Number 12.

S.Amdt. 86 (Inhofe)

Requires a study by the National Academy of Sciences of the effects of the final rule relating to New Source Review promulgated December 31, 2002, and requires an interim report to Congress no later than March 3, 2004. Amendment was agreed to, by a vote of 51-45. Record Vote Number 11. Enacted as Section 356 of H.J.Res. 2 (P.L. 108-7).

Congressional Hearings, Reports, and Documents

U.S. Congress. House. Committee on Energy and Commerce. Subcommittee on Energy and Air Quality. *Comprehensive National Energy Policy.* March 13, 2003. Panel 3, on Fuels, including MTBE and Ethanol Issues.

U.S. Congress. Senate. Committee on Environment and Public Works. Subcommittee on Clean Air, Climate Change, and Nuclear Safety. *CMAQ and Conformity Programs.* March 13, 2003.

U.S. Congress. Senate. Committee on Environment and Public Works. Subcommittee on Clean Air, Climate Change, and Nuclear Safety. *Fuel additives and Renewable Fuels.* March 20, 2003

U.S. Congress. Senate. Committee on Environment and Public Works. Subcommittee on Clean Air, Climate Change, and Nuclear Safety. *Clear Skies Act of 2003, S. 485.* April 8 and May 8, 2003.

Chapter 4

ELECTRICITY RESTRUCTURING: THE IMPLICATIONS FOR AIR QUALITY*

Larry B. Parker and John E. Blodgett

INTRODUCTION

Electricity generation is a major source of air pollution as well as of greenhouse gases. As a result, changes in the electric utility industry raise concern about environmental consequences. Such changes are currently in the offing, both from new generating and transmission technologies and from shifting policy perspectives with respect to competition and regulation. Whether legislation to restructure the industry should include environmental protections has become an issue. Several bills were introduced in the 106th Congress to incorporate such protections into any potential federal restructuring legislation.[1] More legislative proposals are expected to be introduced in the 107th Congress.

This paper reviews the changes now underway in the utility industry from the perspective of their environmental implications — specifically, the potential for electric utility restructuring to increase emissions of air pollutants regulated by the Clean Air Act (CAA) and of greenhouse gases that could be affected by the Kyoto agreement to address climate change.[2] The paper is divided into six parts.

- The *Overview* provides background on the electric utility industry and current restructuring efforts, the industry's emissions, and the fundamental argument with respect to current restructuring activities and potential air pollutant effects. It

---

* Excerpted from CRS Report 98-615 ENR. Updated January 4, 2001.

[1] For a current review of legislation, see Larry Parker and Amy Abel, *Electricity: The Road Toward Restructuring*, Issue Brief IB10006, updated regularly. For a comparison of initiatives introduced in the 106th Congress, see: Larry Parker, *Electricity Restructuring and Air Quality: Comparison of Proposed Legislation*, CRS Report RS20326, updated July 26, 2000.

[2] For current information on the status of the Kyoto agreement and related legislation, see Wayne A. Morrissey and John R. Justus, *Global Climate Change*, Issue Brief IB89005.

identifies electricity demand and air quality regulation as critical to determining air emission impacts from electricity restructuring.

- *The Utility Industry* examines the electricity demand component in more detail. Estimating electricity demand increases in the short- to mid-term and discussing the implications of transmission capacity for interregional electricity transfers, the analysis suggests that restructuring may encourage current trends among utilities to renovate a sizeable amount of existing coal-fired electricity, which generally produces more air pollution and greenhouse gases than alternative kinds of electricity generation.
- *Environmental Regulation* provides background on current air pollution regulations affecting electric utilities and on the resulting complex system of federal and state, pollutant-by-pollutant controls. It discusses regulatory implications for existing capacity and new construction, noting that while the CAA has historically focused on new construction in applying its most stringent standards, several current and prospective regulations would significantly increase controls on existing, coal-fired facilities. It also notes that an increasingly competitive generating market may present significant challenges to the state-directed environmental regimen of the CAA.
- *The Effects of Restructuring and Environmental Actions on Emissions* analyzes the cost-effectiveness of existing coal-fired facilities versus new construction and the environmental effects of increased utilization of existing coal-fired facilities. The analysis indicates that renovating existing coal-fired facilities is generally very cost-effective compared with new, less polluting construction, pointing to a potential for increasing emissions of some air pollutants, especially nitrogen oxides, depending on regulatory actions, and of the greenhouse gas carbon dioxide, which is not regulated.
- *Assessing the Impacts of Restructuring* examines the Federal Energy Regulatory Commission's attempt to estimate the environmental impacts of introducing competition into the wholesale electricity market, and reactions to that analysis. It notes the considerable difficulties in attempting to isolate the potential impact on emissions of restructuring the electricity generation from other technological and policy trends occurring in the industry.
- The *Conclusion* reviews possible responses to potential risks to the environment arising from electricity restructuring. Critical issues are: (1) For conventional air pollutants, whether the existing regulatory regimen will work effectively as the industry structure changes; for some pollutants, such as sulfur oxides, a nationwide emissions "cap" seems secure, but for others, particularly nitrogen oxides, the state-led implementation process may find it difficult to cope with increasingly regional utility industry and environmental challenges. And (2) for greenhouse gases, any controls are contingent on future ratification of the Kyoto Agreement to curtail emissions and on domestic implementation legislation.

OVERVIEW

The Electric Utility Industry and Air Emissions

The industry is massive, with 1996 assets totaling $696 billion, retail sales of $212 billion, and wholesale sales (sales for resale) of $47 billion. It consists of 3,195 utilities — 243 investor-owned, 2010 publicly owned, 932 cooperatives, and 10 federal entities. It is difficult to overestimate the importance of electric service to the nation's economy and individuals' quality of life. In 1996, the average residential customer paid $861 to buy 9,707 kilowatt-hours (809 Kwh monthly) of electricity.[3]

The industry is also a major source of air pollution. The combustion of fossil fuels, which account for 67% of electricity generation, results in the emission of a stream of gases. These gases include several pollutants that directly pose risks to human health and welfare, including sulfur oxides (SO_2), nitrogen oxides (NO_x,), particulate matter (PM), volatile organic compounds (VOCs), carbon monoxide (CO), and various heavy metals, including lead and mercury (Hg). Other gases may pose indirect risks, notably carbon dioxide (CO_2), which may contribute to global warming.[4] (See table 1.)

Of the fossil-fired steam generators, coal-fired facilities contribute a disproportionately large share of these gases. While coal accounts for about 84% of fossil-fuel fired electricity generated, it accounts for 90% or more of the gases listed in table 1 (99% of the Hg). Besides the fuel, the location of a generator can also have important consequences for air pollution impacts (for CO_2, source location is immaterial). Location can be important both with respect to local ambient conditions and, because of long-range transport, to downwind areas. For example, with prevailing air movement from west to east, nonattainment of the ozone air pollution standard in the Northeast has directed attention to the concentration of coal-fired generating facilities in the Midwest as possible contributing sources, particularly of NO_x, which is a precursor of smog-forming ozone, among other effects.

Utilities are currently subject to an array of environmental regulations, which differentially affect both the cost of operating existing generating facilities and of constructing new ones. In particular, air pollution controls impact the construction and operating costs of fossil-fuel fired facilities — hydropower, nuclear, solar, wind, and other nonfossil-fueled fired electricity sources produce essentially zero air pollutants (although they have other environmental impacts). Generally, air quality regulations impose the greatest costs on coal-fired facilities and the least on natural gas-fired ones. This disparity would become greater *if* the U.S. were to accept the greenhouse gas reduction goals of the Kyoto Agreement. Table 2 illustrates the variation among fossil fuels for emissions of NO_x and CO_2.

---

[3] Based on revenues. Statistics from: Energy Information Administration, *Financial Statistics of Major U.S. Investor-Owned Electric Utilities: 1996*, DOE/EIA-0437(96)/1 (Washington, D.C.: December 1997); and Energy Information Administration, *Statistics of Major U.S. Publicly Owned Electric Utilities: 1996*, DOE/EIA-437(96)/2 (Washington, D.C.: March 1998).

[4] Steam-electric utilities produce only minor amounts of VOCs, CO, and lead — on the order of 2% or less of all sources.

The Argument about Restructuring and the Environment

After many decades of operating in a comprehensive, regulated market structure, the electric utility industry is facing significant change, both from new generating and transmission technology and shifting policy perspectives with respect to competition and regulation. At issue is whether these changes will increase air pollution emissions.

Table 1: National Estimated Emissions from Fossil-Fuel, Steam-Electric Utilities — 1996

	CO_2		NO_x		PM_{10}		SO_2		Hg	
	1000 short tons	% all sources	1000 short tons	% all sources	1000 short tons	% all sources	1000 short tons	% all sources	tons %	all sources
Electric Utilities	2,209,287	36	6,103	25	302	11	13,217	67	52	33
Coal	1,911,627		5,395		273		12,426		51.6	
Oil	100,895		208		9		730		0.2	
Gas	195,868		344		1		2			
Other/ Internal Combustion	897		156		19		60			

Sources: CO_2 — DOE, Energy Information Administration, *Electric Power Annual 1998*, Vol. II, p. 42; NO_x, PM_{10}, SO_2 — EPA, *National Air Quality and Emissions Trends Report, 1998* EPA 454/R-00-003 (March 2000), Tables A-4, A-6, and A-8 [http://www.epa.gov/oar/aqtrnd98/fr_table.html]; Hg — EPA, *Mercury Study Report to Congress*, Vol. 1, "Executive Summary" EPA-452/R-97-003 (December 1997), p. 3-6 [Hg data estimated annual emissions 1994/1995].

Table 2: NO_x and CO_2 Emission Rates by Fuel Source

Fuel	NO_x Emissions (lb./mmBtu)	CO_2 Emissions (lbs. carbon/mmBtu)
Coal	0.1 - >2	55.9
Natural Gas	0.005 - >1	31.7
Residual Fuel Oil	0.05 - >1	46.8

Range for NO_x reflects the difference between best available control technology and emissions from an uncontrolled existing power plant.

Sources: NO_x — Larry Parker, *Nitrogen Oxides and Electric Utilities: Revising the NSPS*, CRS Report 96-737, July 25, 1997; CO_2 — EIA, *Emissions of Greenhouse Gases in the United States: 1987-1992* (Washington, D.C., 1994), Appendix A; and EIA, *Emissions of Greenhouse Gases in the United States: 1987-1994* (Washington, D.C., 1995), p. 18.

Technology

The advent of new generating technologies, particularly natural gas-fired combined cycle, has both lowered entry barriers to competitors of traditional utilities and lowered the marginal costs of those competitors below that of some traditional utilities. As noted by the Federal Energy Regulatory Commission (FERC), smaller and more efficient natural gas-fired, combined cycle generation plants can produce power on the grid for between 3 cents and 5 cents per kilowatt-hour (Kwh). This is typically less than for the larger coal-fired (4-7 cents/Kwh) or nuclear (9-15 cents/Kwh) plants built by traditional utilities over the past decade.[5] Indeed, it is less than the average generating costs of some utilities. Coupled with advances in generating technology have been advances in transmission technology that permit long distance transmission economically and permit increasingly coordinated operations and reduced reserve margins.

This technological advancement has been combined with legislative initiatives, such as the Energy Policy Act of 1992 (EPACT), to encourage the introduction of competitive forces into the electric generating sector. This shift in policy continues with the promulgation of FERC Order 888 encouraging competition in the wholesale electricity market and implementation by some states of retail competition initiatives.

Restructuring

The policy shift underlying the changes occurring in the electric utility industry is a growing belief that the rationale for the current economic regulation of electric utilities at both the federal and state levels — that electric utilities are natural monopolies — is being overtaken by events, and that market forces can and should replace some of the current regulatory structure. Regulation and rate-of-return ratemaking[6] arguably exist as a partial substitute for the marketplace. The emerging trend in the industry suggests that regulation is an imperfect substitute for the marketplace and that with emerging new generating and transmission technologies, real self-regulating market forces are now able to replace government regulation in many instances. This substitution could result in a more efficient allocation of the country's resources and provide consumers with more accurate price signals regarding the actual cost of electricity.

The restructuring effort attempts to reduce and alter the role of government in electric utility regulation by identifying transactions, industry segments, regions, or specific activities that might no longer be the subject of economic regulation. Current proposals to increase competition in the electric utility industry involve segmenting electric functions (generation, transmission, distribution) that are currently integrated (or bundled) in most cases (both in terms of corporate and rate structures). The overall purpose of restructuring is to promote economic efficiency, which will presumably lead to lower overall rates.

Some argue that this singular focus on economic efficiency could come at the expense of other values that the regulatory system traditionally has balanced against economic efficiency, particularly equity and environmental considerations. The environmental concern with respect to restructuring is that the new economic signals being given by a competitive generation market could result in increased emissions of undesirable pollutants for two basic reasons: (1) lower baseload prices resulting from restructuring would increase electricity demand and, therefore, increase generation and emissions; and (2) the restructured generating market's revaluation of existing facilities to the marginal cost of constructing new capacity (along with their low operating costs) would encourage the rehabilitation and full utilization of these older, more polluting generating facilities.

5 FERC, "Promoting Wholesale Competition ... [Final Rule]," 61 *Federal Register* (May 10, 1996), 21544.

6 Rate-of-return ratemaking means that a regulatory body allows the utility to obtain a guaranteed rate of return on investment. The regulatory body specifies a utility's legitimate costs and approves rates that allow the recovery of those costs plus a regulatorily determined acceptable profit. Wholesale sales are regulated by FERC; retail sales are regulated by state Public Utilities Commissions, which may also regulate investment and debt.

Proponents of restructuring argue that it would increase efficiency and reduce electricity costs. To the extent greater competition and lower costs translate into lower prices, demand can be expected to rise (and incentives to conserve electricity and for new technologies such as renewable energy can be expected to decline). More demand would require more generation, resulting in more emissions. How much emissions might increase would depend on what facilities generate the additional power and on controls imposed by existing or prospective Clean Air Act requirements, as discussed below. Some cost studies indicate that the lion's share of cost savings from restructuring would come from the increased use of existing coal-fired capacity[7] — which is disproportionately more polluting than alternative sources of power. If true, then the need for new (cleaner) generating capacity could be delayed by restructuring, as production from existing capacity is maximized.

Based on the above, the general scenario goes as follows. A competitive generating sector would result in a revaluation of generating assets — i.e., moving from a traditional embedded-cost valuation scheme to a market valuation scheme, which would increase the value of some generating capacity and decrease the value of other generating capacity. Competition would tend to move the value of generating capacity to the marginal cost of constructing new capacity, generally represented at the current time by a new natural gas-fired, combined-cycle facility. In general, older facilities that have been fully depreciated would tend to have market values greater than their current book value under regulation; in contrast, newer, capital intensive facilities (such as some nuclear plants) would have market values less than their current book value. (Case-by-case valuation would be affected by location, availability of alternatives, and electricity demand.) In addition, the Clean Air Act typically imposes its most stringent pollution controls on new powerplant construction, permitting existing capacity to meet less stringent and less costly standards. This differential impact may give some older facilities a competitive operating cost advantage to complement their low, depreciated cost basis.

The new valuation, combined with low operating costs, would encourage operators to maximize generation from their existing facilities. The trend toward increased utilization have already begun. In 1995, coal-fired facilities operated at a 62% capacity factor. By 1999, operation of coal-fired capacity had increased to 67%. The upper limit here is unclear — the economic and environmental advantages of new technology, such as natural gas-fired, combined-cycle technology (a very clean technology) may be sufficient in some cases to overcome the advantages of expanding use of existing plants.

Environmental Implications

It is this renewed attractiveness of existing capacity under restructuring, specifically of coal-fired capacity, along with the potential that demand for electricity may rise (and energy conservation slacken) if prices decline, that raises environmental concerns. Absent effective controls, burning more coal will produce more emissions than alternative sources of electricity generation — and much of that coal capacity is in the Midwest, which is currently a center of attention for reducing NO_x emissions.

[7] For example, see Michael T. Maloney and Robert E. McCormick, *Customer Choice, Consumer Value: An Analysis of Retail Competition in America's Electric Industry*, prepared for the Citizens for a Sound Economy Foundation (1996).

Except for CO_2, the regulatory regimen of the Clean Air Act provides authorities for controlling the potential increase in emissions — assuming they are effectively implemented. Existing controls "cap" SO_2 emissions in the 48 contiguous states and the District of Columbia, and there is no reason to question the effectiveness of the cap in the future, regardless of the changes underway in the utility industry. For NO_x emissions, control and implementation is more complicated, primarily because implementation of much of the process lies with the states. Any increase in NO_x emissions in the Midwest could complicate an already difficult process underway to reduce the region's NO_x emissions, which contribute to ozone nonattainment in the Northeast.[8] How this regional, state-implemented process would be affected by restructuring is not certain.

CO_2 is not currently regulated. Any increase in fossil fuel-fired generation will increase CO_2 emissions, with coal producing about 75% more carbon emissions than natural gas on a Btu basis. *If* the U.S. were to ratify the Kyoto Agreement, which would require the U.S. to reduce greenhouse gas emissions to below 1990 levels, any increases would have to be rolled back or offset.[9] The effort required would be increased if restructuring differentially advantaged coal.

Ultimately, whether developments in electricity generation and demand lead to increased emissions of air pollutants depends on the implementation of the CAA (and on any new requirements that might be enacted); while for CO_2, increases are likely unless Congress ratifies the Kyoto Agreement and enacts implementing legislation (an uncertain prospect). Those who are focused on preventing environmental deterioration tend to take a precautionary stance, to propose immediate preventative measures, and to argue that such measures be attached to available legislative vehicles. In contrast, those who doubt that there will be significant environmental effects and/or who are focused on the substantial regulatory structure in place tend to take a wait-and-see position.

The current attention on increased emissions from coal-fired generation may address the clearest and most quantifiable risk to the environment from restructuring, but with so many changes underway, the ultimate outcome remains uncertain. Some trends are already manifest, such as renovation of existing coal-fired capacity. Others are just emerging, such as a "green market" in California, in which consumers can take into account environmental costs in their purchasing decisions. Some effects remain to be determined in the future, such as the implications of new price signals for demand and conservation; the implication of new cost valuations for the choice of new generating technologies; developments in transmission capacity; and the effectiveness of ongoing environmental programs. These complexities and their interactions are explored in more detail in the following discussions.

THE UTILITY INDUSTRY

Utility industry variables affecting emissions include: overall demand for electricity, which will respond to any changes in prices; the mix of fuels, which will be strongly affected by demand, especially for baseload capacity; and transmission capacity, which will affect

[8] For a discussion of those efforts, see Larry Parker and John Blodgett, *Air Quality: EPA'sOzone Transport Rule, OTAG, and Section 126 Petitions — A Hazy Situation?* CRS Report 98-236, updated July 14, 2000.

[9] For a discussion of U.S. global climate change policy, see Larry Parker and John Blodgett, *Global Climate Change Policy: From "No Regrets" to S. Res. 98*, CRS Report RL30024, January 12, 1999.

what generators can respond to demand. Also crucial are environmental regulations that set limits on certain emissions and/or shift costs among generating facilities. This interactive matrix makes it difficult to separate out the environmental effects of any one component, such as restructuring.

Meeting Future Electricity Demand

In general, the United States has more electric generating capacity than it needs to maintain reliability. Currently, capacity margins[10] of between 12% and 17% are considered necessary to maintain adequate reliability.[11] Nationwide, U.S. capacity margins average 15% — varying from about 13% to 18% on a regional basis.[12] These capacity margins are expected to fall in the future as demand increases. The planned capacity margin in 2008 is 9.1%, unless announced new merchant plant capacity comes on line as intended. In that case, the 2008 capacity margin would be 15.6%.

On the surface, these numbers would suggest that there would be a general need for new capacity in the short- to mid-term (5-10 years), providing opportunities for different generating technologies, such as natural gas combined-cycle technology, coal-fired technologies, renewables, and nuclear power. However, this may not be the case for some regions. Much of the planned construction to meet the capacity growth needs identified above is designed to meet anticipated peak load, not baseload needs.[13] Capacity that is not dispatchable on demand, such as some renewables and nuclear power, may not fit the demand curve over this time period. For example, utilities representing the southeastern U.S. estimate that nearly 90% of the projected 26,990 Mw of new capacity coming on line over the next 10 years will be non-baseload capacity. Similarly, the utilities representing the industrial Midwest estimate that 94% of the projected 13,500 MW of new capacity coming on line will be combustion turbines (a technology typically used for meeting peak load).[14]

This lack of planned construction for new baseload generating units reflects, in part, an existing surplus of baseload capacity, particularly coal-fired capacity.[15] In 1995, coal-fired capacity operated at a 62% capacity factor. By 1999, this had increased to 67%.[16] If demand and economics justified it, this average could improve to 75% or more. An increase to 75%

---

[10] Capacity margins should not be confused with reserve margins. Capacity margin is the difference between generating capacity and peak load expressed as a percent of generating capacity. Reserve margin is the difference between generating capacity and peak load, expressed as a percent of peak load. Thus, a 17% capacity margin is roughly equivalent to a 20% reserve margin.

[11] Capacity margins are generally set according to a Loss of Load Probability (LOLP) calculation — a measure of the long-term expectation that a utility will be unable to meet demand. A 1 day in 10 year LOLP is typical.

[12] Data for 1999. North American Electric Reliability Council, *Reliability Assessment: 1999-2008* (Princeton, NJ: NERC, May, 2000), p. 14.

[13] Baseload refers to the minimum amount of electric power delivered or required over a given period at a constant rate. Baseload powerplants, like nuclear plants, are designed to operated whenever they are available (generally over 60% of the time).

[14] North American Electric Reliability Council, *Reliability Assessment: 1996-2005* (Princeton, NJ: NERC, October 1996).

[15] The lack of planning also reflects the shortening of lead-times for new construction, uncertainty about future demand, and uncertainty about the future structure of the generating sector.

[16] It is this trend in coal-fired generation utilization that caught the attention of EPA and the possibility for action under the New Source Review requirements of the Clean Air Act. For more information, see Larry B. Parker and John E. Blodgett, *Air Quality and Electricity: Enforcing New Source Review*, CRS Report RL30432 (January 31, 2000).

capacity would be equivalent to about 23,000 Mw of baseload capacity — sufficient to meet increases in aggregate baseload demand for a couple of years, depending on transmission capacity constraints. (An increase to 85% capacity would be equivalent to about 53,000 Mw.) Thus, it would appear that under current expectations, existing baseload facilities, such as nuclear plants, and new baseload construction, such as natural gas combined-cycle, may in many cases be competing against existing coal-fired facilities for the next 5-10 years.

Transmission Capacity

The degree to which existing coal-fired capacity competes against other baseload technologies will be partially dependent on transmission capacity. Under ideal economic conditions, the price of providing baseload electricity would tend to levelize across the country, reflecting a nationwide market for such electricity. In reality, this is unlikely to occur until and unless substantial improvements are made in transmission capacity and the robustness of the transmission grid. An increase in market forces in the generating sector does not necessarily translate into the increased transmission capacity and robustness that would allow consumers to fully exploit potential generation savings.

Under current restructuring proposals, the transmission sector remains a monopoly controlled by rate-of-return regulation. The history of this approach to transmission planning has resulted in a system focused on and justified by local reliability concerns, not a system concerned with maximizing economic efficiency on a nationwide or even interregional basis. How well and how completely the regulatory structure can be changed to facilitate the dynamics of a deregulated generating sector is difficult to predict. Market prices may regionalize, reflecting the increasingly regional control of transmission, but large-scale interregional transactions may be several years away.

If transmission barriers result in largely regional markets, marginal costs for baseload capacity may differ between regions. For example, regions with substantial excess coal-fired capacity may have low marginal costs based on the incremental costs of increased capacity utilization. Other regions, with substantial increasing demand, may have marginal costs based on new construction costs, such as building a natural gas-fired combined-cycle plant or a coal-fired fluidized bed combustor. Depending on price, a generating technology that is competitive "on average" may not be competitive within a specific region, because of low-cost alternatives; likewise, a "higher cost" generating technology that is non-competitive "on average" may be competitive within a specific region because of the higher cost of alternatives.

Implications of Utility Developments

All of these factors will be summed up in the price for baseload power. It is generally assumed that deregulation of the generating sector will encourage the development of marginal cost pricing.[17] In particular, deregulation will clearly expose the substantial cost

[17] Marginal cost has been used by some public utility commissions to determine appropriate rates between different customer classes for several years, and utilities have also experimented with "time-of-day" rates that reflect

differences between baseload generation and peak generation. While baseload facilities generally run at over 60 percent capacity, peak demand facilities run at under 20 percent capacity. This substantial difference in utilization, among other differences, means that peak power will cost more under restructuring than it does now, when the cost is generally rolled in with the less expensive baseload power.[18]

At least in the short-term, this stratification of electricity pricing may mean that the market price for baseload power will be considerably lower than the current average electricity price would indicate. This would encourage the use of existing baseload capacity with low operating costs (e.g., coal-fired capacity) and discourage constructing new baseload facilities, particularly those technologies requiring substantial investment (e.g., nuclear power). Low baseload prices may also discourage development of non-dispatchable power sources (e.g., some renewable technologies) and installation of some conservation technologies. Higher prices for peaking power would encourage technologies designed for such load (e.g., combustion turbines), and technologies designed to reduce such loads (e.g., load management techniques). In the long term, if prices for electricity decline, electricity use is likely to increase and incentives to conserve electricity are likely to decrease. Long-term declining prices could also reduce incentives for new technologies, including some renewable energy technologies.

How these different effects play out will determine the potential for increased emissions from restructuring. Although the overall effect on emissions is difficult to assess, involving several currently unquantifiable variables, the most substantial environmental effect in the short- to mid-term is likely to come from enhanced operation of existing coal-fired capacity. Whether one can ascribe that effect strictly to restructuring is debatable, however.

Environmental Regulation

The Clean Air Act imposes a complex regulatory structure on air pollution sources. From an historical perspective, the regulatory environment for a major emission source, like an electric generating facility, has been largely dependent on two factors: (1) Where the facility is located (in an area meeting clean air standards, or in an area not attaining them) and, (2) How old the facility is (new or old source). Other factors, such as facility size and specific pollutants controlled, feed off these two factors. This framework is changing, however, as illustrated in the following case study on NO_x.

Example: Nitrogen Oxide Control

Nitrogen oxides, both directly and because they contribute to formation of ozone, raise human health and environmental concerns that bring them under the purview of the CAA.

---

marginal costs across time. Under restructuring, generating costs may move more in the direction of "time of day" pricing as more reflective of actual costs than the current average cost method.

[18] As stated by a study done for the American Gas Association study of future electric generation: "In principle, retail deregulation and retail wheeling, should radically change the current pricing structure for end-use electricity. Peak pricing will increase sharply and off-peak pricing will decrease sharply." Harry Chernoff, *Existing and Future Electric Generation: Implication for Natural Gas*, Study prepared for the American Gas Association, Policy Analysis Group, by Science Applications International Corporation (October 1996) p. 23.

Nitrogen dioxide (NO_2), the index compound for nitrogen oxides, can irritate the lungs and lower human resistance to various respiratory infections, such as influenza. In combination with volatile organic compounds (VOCs) and in the presence of heat and sunlight, NO_x forms ozone, for which human health concerns include lung damage, chest pain, coughing, nausea, throat irritation, and congestion. Ozone also exacerbates the effects of bronchitis, heart disease, emphysema, and asthma.[19] In addition, nitrogen oxides contribute to the formation of fine particulates, suspected of significant human mortality and morbidity effects and for which EPA recently set new standards that will become effective in 10 to 15 years.[20]

Environmental concerns about NO_x emissions include its transformation into nitric acid, a component of acid precipitation; visibility impairment; and adverse effects of ozone on plant life.[21] In addition, EPA estimates that up to 40% of the nitrogen "loading" in the Chesapeake Bay, resulting in excessive nutrient enrichment, is the result of deposition of air-borne nitrogen oxides. In the West, nitrogen oxides contribute to visibility impairment, particularly in southern California. These multiple effects result in multiple control measures under the Clean Air Act, as described below.

Air Quality Regulations Impacting on Utilities

Primary **National Ambient Air Quality Standards (NAAQS)** set maximum levels of permitted pollution concentrations nationwide. NAAQS are federally enforceable with specific deadlines for compliance; they are required by Section 109 of the CAA to protect the public health with an "adequate margin of safety." They are periodically reviewed to take into account the most recent health data. Three NAAQS may result in NO_x controls: NAAQS for nitrogen dioxide, ozone, and fine particulates.

In 1994, all monitoring locations in the U.S. were in compliance with the NO_2 NAAQS; however, compliance with the ozone NAAQS remains elusive in several parts of the country, particularly in southern California, the Texas Gulf Cost, and the Northeast corridor (from Virginia to Maine). Because NO_x is a precursor to ozone formation, NO_x control represents an important component in reducing ozone pollution. In recognition of the multi-state nature of the ozone problem in the Northeast, the 1990 CAA Amendments created an Ozone Transport Commission (OTC) to development and coordinate emission reduction efforts for the area. In addition, in 1998, the EPA promulgated a new ozone transport rule that would control NO_x emissions for 21 eastern states, and ten states petitioned the EPA to control NO_x emissions in the Midwest under section 126 of the CAA.[22]

---

[19] For a discussion of human health effects of air pollution, see Morton Lippmann, "Health Benefits from Controlling Exposure to Criteria Air Pollutants," in John Blodgett, ed., *Health Benefits of Air Pollution Control: A Discussion*, CRS Report 89-161, February 27, 1989, pp. 75-144.

[20] John Blodgett, et al., *Air Quality Standards: EPA's Final Ozone and Particulate Matter Standards*, CRS Report 97-721 (Updated June 19, 1998).

[21] For a discussion of ozone and acid precipitation effects on vegetation, see David S. Shriner, et. al., *Response of Vegetation to Atmospheric Deposition and Air Pollution: State of Science and Technology Report 18* (Washington, D.C.: National Acid Precipitation Assessment Program, December 1990).

[22] For more information, see Larry Parker and John Blodgett, *Air Quality: EPA's Ozone Transport Rule, OTAG, and Section 126 Petitions — A Hazy Situation?* CRS Report 98-236, updated July 14, 2000. For recent activities with respect to these initiatives, see: Larry B. Parker and John E. Blodgett, *Air Quality and Electricity: Initiatives to Increase Pollution Controls*, CRS Report RS20553, December 28, 2000.

For areas in attainment with these NAAQS, the CAA mandates states to require new sources, such as powerplants, to install **Best Available Control Technology (BACT)** as the minimum level of NO_x control required of a new powerplant.[23] State permitting agencies determine BACT on a case-by-case basis, taking into account energy, environmental and economic impacts. BACT can be much more stringent than the federal New Source Performance Standard (NSPS — described below), but can not be less stringent than NSPS. Existing sources are not required to install controls in attainment areas.

For areas *not* in attainment with one or more of these NAAQS, the CAA mandates states to require new sources to install **Lowest Achievable Emissions Rate (LAER)** technology. Along with offset rules, LAER ensures that overall emissions do not increase as a result of a new plant's operation. LAER is based on the most stringent emission rate of any state implementation plan or achieved in practice without regard to cost or energy use. It may not be less stringent than NSPS. Existing sources are required to install **Reasonably Available Control Technology (RACT)**, a state determination based on federal guidelines.

A **Prevention of Significant Deterioration (PSD)** program (Part C of the CAA) focuses on ambient concentrations of pollutants (including NO_2) in "clean" air areas of the country (i.e., areas where air quality is better than the NAAQS). The provision allows some increase in clean areas' pollution concentrations depending on their classification. In general, historic or recreation areas (e.g., national parks) are classified class 1 with very little degradation allowed while most other areas are classified class 2 with moderate degradation allowed. Class 3 areas are permitted to degrade up to the NAAQS. New sources in PSD areas must undergo preconstruction review and must install BACT; state permitting agencies determine BACT on a case-by-case basis, taking into account energy, environmental, and economic impacts. More stringent controls can be required if modeling indicates that BACT is insufficient to avoid violating PSD emission limitations, or the NAAQS itself.

A complement to the PSD program for existing sources is the **regional haze** program (section 169A) that focuses on "prevention of any future, and the remedying of any existing, impairment of visibility" resulting from manmade air pollution in national parks and wilderness areas.[24] Among the pollutants that impair visibility are sulfates, organic matter, and nitrates. In 1999, the EPA promulgated a regional haze program, which, would entail more stringent controls on NO_x and SO_2. However, like the fine particulate NAAQS, it will be several years before any regional haze program might result in controls.

New Source Performance Standards (NSPS) are federal standards defining the minimum controls necessary for new sources regardless of their location — in contrast to the PSD and NAAQS standards that focus on ambient concentrations of pollutants. EPA's NSPS determinations represent the floor for state BACT and LAER determinations in case-by-case situations.

Required under Section 111 of the CAA, NSPS require major new sources to install the best system of continuous emission reduction which has been adequately demonstrated. In making such an assessment, the CAA requires EPA to take into account "the cost of achieving such reduction and any nonair quality health and environmental impact and energy requirements." To keep controls abreast of technological innovations, the CAA originally

[23] More stringent controls can be required if modeling indicates that BACT is insufficient to avoid violating the NAAQS.

[24] See James McCarthy, et al., *Regional Haze: EPA's Proposal to Improve Visibility in National Parks and Wilderness Areas*, CRS Report 97-1010, updated July 9, 1998.

required EPA to review and revise NSPS every four years. But at the time of enactment of the 1990 CAA Amendments, the last revision of the NO_x NSPS for electric and non-electric steam generating units had occurred in 1979. With substantial technological improvements in controlling NO_x having occurred during the 1980s, the 1990 Amendments (title IV, section 407(c)) required EPA to promulgate a new NOx NSPS for electric and non-electric steam generating units by 1994 — a deadline EPA did not meet. In September, 1998, EPA did promulgated a new NO_x NSPS. It is considerably more stringent than the 1979 standard for coal-fired facilities, but not particularly stringent for natural gas or oil-fired facilities.[25]

The **acid deposition control** provisions of title IV of the 1990 Amendments focus on total emissions from existing sources of sulfur dioxide and nitrogen oxides. For nitrogen oxides, section 407 of title IV requires tangential- and wall-fired (dry bottom, not cell burner equipped) boilers (group 1 boilers) designated to meet 1995 phase 1 reductions to meet an emission limitation based on low- NO_x burner technology. Regulations for phase 1 NO_x reductions were finalized in 1995. For phase 2 in the year 2000, remaining group 1 boilers are required to meet the same standard (or more stringent if technology and costs permit) as those covered in phase 1, and boilers with other firing configurations (group 2 boilers) are required to meet standards based on available technology that is comparable in cost to low-NO_x burners. EPA finalized regulations for phase 2 group 1 and group 2 boilers in 1996 (61 *Federal Register* 245, pp. 67112-67164).

Implications of Air Quality Regulations for Utilities

In the light of changes in the utility industry, this mix of air quality regulations has important consequences for (1) utilities' choices both for construction of new facilities and operation of existing ones, and (2) the potential effectiveness by which federal and state air pollution controls apply to a changing industry structure.

New Construction and Existing Sources

For constructing new powerplants, the CAA envisions the federal NSPS and the state PSD/BACT program as the baseline for control efforts in attainment areas, and the state-set LAER and federally-based offset requirements as the baseline in nonattainment areas. For SO_2, federal offset requirements overlay these other requirements. The costs of installing NSPS (or BACT or LAER or obtaining offsets) on new construction fall most sharply on coal-fired facilities. This could disadvantage coal in choices among technologies for new generation.

At the same time, the historically less stringent controls on existing coal-fired facilities — none in attainment areas, RACT in nonattainment areas — have clearly advantaged existing sources, particularly coal, in competing with new sources for meeting generating needs. This may be changing. For existing facilities, especially coal-fired facilities, a host of new regulatory initiatives may result in more stringent controls for a number of possible pollutants.

In addition, in what could crucially affect the potential costs of reconditioning and extending the life of existing coal-fired plants, EPA, together with the Department of Justice,

[25] See Larry Parker, *Nitrogen Oxides and Electric Utilities: Revising the NSPS*, CRS Report 96-737, updated

has initiated a New Source Review (NSR) enforcement process to reduce pollution from existing sources. The first overt action under this process occurred November 3, 1999, when the Justice Department filed seven lawsuits against electric utilities in the Midwest and South, charging them with violations of the NSR requirements of the CAA. EPA also issued an administrative order against the Tennessee Valley Authority, alleging similar violations.

The crux of the enforcement actions is the "preconstruction" permitting process of the NSR, which is designed to ensure that newly constructed facilities, or substantially modified existing ones, do not result in violations of applicable air quality standards. The question the enforcement actions raise is whether the specified facilities engaged in rehabilitation actions that represent "major modifications" of the plants, in which case the CAA would require the installation of best available control technology – BACT.

The crucial definition of "major modification" derives from an EPA ruling that a life extension project by Wisconsin Electric Power Company (WEPCO) triggered NSR requirements. Since 1992, after considerable litigation and congressional debate, the "test" to determine the applicability of NSR compares whether a facility's projected actual emissions after the modification are more than its actual emissions before the modification. Utilities argue that the "modifications" EPA cites in the suits were just routine maintenance, which does not trigger NSR. If EPA's position in these suits is upheld, this could have the dual effects of increasing the costs to utilities of expanding the use of coal-fired utilities in the future and of reducing the emissions from coal-fired facilities.[26]

Table 3 identifies the range of environmental actions that are beginning to affect or may in the future affect emissions from fossil-fuel fired facilities. As discussed in the next section, these controls could have a substantial influence on the cost of power from coal-fired facilities, making them less attractive in a competitive marketplace.

Table 3: Potential Control on Existing Sources

Pollutant	Potential Controls on Existing Sources
Nitrogen Oxides	Title IV, sec. 407 Ozone Transport Commission (OTC) Rules Ozone Transport Rule Section 126 Petitions Revised Ozone NAAQS Fine Particulate NAAQS New Source Review Enforcement Regional Haze Rule More stringent Legislation[a]
Sulfur Oxides	Title IV Fine Particulate NAAQS New Source Review Enforcement Regional Haze Rule More stringent Legislation[a]

October 13, 1998.

[26] On the NSR enforcement actions, see Larry B. Parker and John E. Blodgett, *Air Quality and Electricity: Enforcing New Source Review*, CRS Report RL30432, Jan. 31, 2000.

Mercury	Potential EPA regulation as a HAP NE Action Plan on Mercury Potential Legislation[a]
Carbon Dioxide	Potential ratification of Kyoto Agreement Potential Legislation[a]

[a] For information on current legislative proposals relating restructuring to environmental controls, see Larry Parker and Amy Abel, *Electricity: The Road Toward Restructuring*, CRS Issue Brief IB10006. For a review of legislation introduced in the 106th Congress, see: Larry Parker, *Electricity Restructuring: Comparison of Comprehensive Bills*, CRS Report RL30087, July 24, 2000; and, Larry Parker, *Electricity Restructuring and Air Quality: Comparison of Proposed Legislation*, CRS Report RS20326, July 26, 2000.

IMPLEMENTATION UNDER RESTRUCTURING

The mix of regulatory authorities results in a complex federal-state process for regulating the industry. The state-regulated utility system meshed reasonably well with the state-implemented air quality controls. As the utility industry becomes more competitive and potentially more regional, and as air quality problems also become more regional (regional haze, longrange pollutant transport), state-directed controls on existing sources may prove less efficient and effective than previously.

These regional challenges may reprise the past inability of the state-led process to control acid rain, the result of long range transport of SO_2, for which utilities are a major source. As a result of the failure of the state-based process to address this problem, Congress in 1990 added the acid rain program to the CAA, which established a national "cap" on emissions. This has proven an efficient program; SO_2 credits are not excessively expensive and the most popular technology for new construction — natural gas-fired combined-cycle technology — produces almost no SO_2 emissions.[27] The flexibility and straight-forward compliance mechanism of this "cap and trade" program would seem to mesh well with a flexible, competitive utility industry, so electricity restructuring would not appear to create any serious implementation problems for this SO_2 control program.

However, the acid rain program is discrete; there is no comparable nation-wide "cap and trade" program for other pollutants. For example, as noted earlier, the 1990 CAA Amendments did create an Ozone Transport Region in ten northeastern states for addressing ozone transport (and NO_x, as a precursor) and it authorized EPA to create others. However, this effort may be inadequate to bring the Northeast into compliance with the Ozone NAAQS. To bring under control additional sources of long-range transport, EPA created a 21-state Midwest-Northeast region (where substantial coal-fired NO_x emission increases could occur) subject to the promulgated Ozone Transport Rule, a feature of which is a voluntary, state-implemented NO_x "cap and trade" program. However, EPA does not have authority to require the states within the region to act in concert or to impose uniform rules for cap and trade as in the acid rain program. Thus, how this regional effort to control NO_x would work in practice remains to be seen; interstate disagreements have surfaced, and industry restructuring could change emissions patterns in ways that could exacerbate them.[28]

[27] U.S. Environmental Protection Agency, *1996 Compliance Report, Acid Rain Program*, EPA 430-R-97-025 (June 1997).

[28] For recent actions with respect to the Ozone Transport Rule, see: Larry B. Parker and John E. Blodgett, *Air Quality and Electricity: Initiatives to Increase Pollution Control*, CRS Report RS20553, December 28, 2000.

The potential for diverse state requirements in the region could lead to inconsistent requirements that could pose barriers to restructuring the industry — or opportunities. Differing requirements could allow utilities to choose which state had the least stringent requirements, while the power could be transmitted to the location of demand; or inconsistent requirements — or uncertain ones — might be an added incentive for construction of generating capacity that is clean and hence not subject to them.

The regulatory dynamic of the Clean Air Act has no direct consequence for potential increases in CO_2 emissions under utility restructuring: CO_2 is not subject to CAA regulation and any controls are prospective, contingent on U.S. ratification of the Kyoto agreement and on domestic implementing legislation. The uncertainty of legislative action on CO_2 compared to the potential for action on restructuring legislation in the next couple of years has led some to find in the restructuring issue a surrogate for a debate on CO_2 controls and global climate change in general. This situation adds complexity to the restructuring debate.

The Effects of Restructuring and Environmental Actions on Emissions

As suggested previously, restructuring involves the interplay of many factors affecting emissions. As indicated, some, such as renovating existing coal-fired capacity, represent a furthering of an existing trend. Others, such as green pricing, represent a new trend created by restructuring. Although the overall effect on emissions is difficult to assess, involving several currently unquantifiable variables, the most substantial environmental effect in the short- to mid-term arises from the potential for enhanced operation of existing coal-fired capacity. However, how much one should ascribe that effect to general trends in the industry vis a vis restructuring is debatable.

Economics and Coal-Fired Generation

A general trend in the electric utility industry for over a decade has been the renovation of existing capacity beyond its initial lifespan (especially coal-fired capacity) in lieu of constructing new capacity. If restructuring results in a stratification of electricity pricing in terms of baseload, intermediate, and peak power, the low price of baseload capacity could provide additional impetus to refurbish existing coal-fired capacity and to maximize operation of such power. Likewise, lower baseload prices would likely reduce incentives to conserve electricity and to develop new non-peaking technologies, including renewable energy. As noted above, substantial amounts of underused coal-fired capacity currently exist. The degree to which it is competitive over the next 5-10 years will depend primarily on two factors — cost of enhanced maintenance to extend the life of the facilities (life extension), and potential for additional pollution control costs (which is discussed in the next section).

Depending on the condition of an existing coal-fired facility, reconditioning can be a very economic means of adding baseload capacity.[29] This reconditioning process, called life

[29] ICF Incorporated, *Repowering and Life Extension: Background Paper*, prepared for the Office of Atmospheric Programs and Office of Air Quality Planning and Standards, EPA (draft report) (February 1995).

extension, can help halt and partially reverse the deterioration of a power plant's efficiency and reliability during continued operation. Over time, the operation and maintenance (O&M) of a powerplant increases, along with its heat rate. For example, based on FERC data, EPA assumes the median O&M costs for coal-fired facilities up to 10 years old is $17.60/Kw , compared with the median costs for a facility more than 30 years old of $31.20/Kw.[30] However, EPA believes that much of this increase (about $9.40/Kw) represents continuous reconditioning efforts to extend the life of the plant — that "life extension" efforts increasingly represent a continuing upgrading process, rather than a one-time reconstruction of the power plant.[31]

Thus, one-time projections for life-extension costs overestimate the incremental cost of this effort. EPA estimates that the cost to extend power plant life from 40 to 65 years will be on the order of $8.8/Kw per year in additional O&M costs — or 1.4 mills/Kwh. Assuming the power plant has been well-maintained up to now, this cost would appear quite attractive for an additional 20-25 years of operation. Including estimated O&M and fuel costs, such power plants would generate electricity for about 2 cents/Kwh.[32] In general, the potential for rising fuel prices is considered small in the case of coal. There appears to be ample supply of coal available at current prices.

Against existing coal-fired capacity is newly constructed natural gas combined-cycle technology. Conventional wisdom within the industry is that, based on current trends in generating technology and fuel costs, the technology of choice for new construction will be natural gas-fired combined-cycle plants. To illustrate the sensitivity of new natural gas-fired facilities to fuel costs and technology improvements, CRS analyzed four different cases. The results are presented in table 4. CRS estimates the annual costs on a levelized basis for a natural gas combined-cycle plant at about 2.4-2.5 cents/Kwh, with costs rising to 3.4-3.5 cents/Kwh if natural gas prices rise to $3.50/mmBtu compared with $2.25/mmBtu assumed in the base-case calculations. While very competitive for new construction, it is not quite competitive, *in general*, to renovating existing coal-fired capacity.[33]

Table 4. Costs of New Natural Gas-fired Combined-cycle Facility (1995 dollars)

	Base case	**High fuel cost case**	**Higher efficiency base case**	**Higher efficiency/high fuel cost case**
Efficiency Assumption	7,300 Btu/Kwh	7,300 Btu/Kwh	6,800 Btu/Kwh	6,800 Btu/Kwh
Fuel Cost Assumption	$2.25/mmBtu	$3.50/mmBtu	$2.25/mmBtu	$3.50/mmBtu

---

30 U.S. Environmental Protection Agency, Office of Air and Radiation, *Analyzing Electric Power Generation Under the CAAA* (July 1996), p. A3-11. Estimates in 1995 $.

31 Part of this represents a strategy by utilities to avoid having to comply with New Source Performance Standards (NSPS) at their existing facilities by not triggering the WEPCO rule, which requires existing facilities to achieve NSPS under some circumstances. It is this strategy that EPA and Department of Justice are attacking with the NSR enforcement actions discussed previously.

32 Calculation assumes heat rate of 10,000 Btu/Kwh, fixed O&M costs of $31 Kw/yr (not including incremental life extension costs discussed in the text), fuel costs of $1.30 Btu/Kwh, and a 70% capacity factor.

33 For reference, CRS calculates that a new coal-fired steam generator would produce electricity for about 3.5 cents/Kwh, confirming the conventional wisdom with respect to natural gas.

Total Costs	2.5 cents/Kwh	3.4 cents/Kwh	2.4 cents/Kwh	3.3 cents/Kwh

Other assumptions include capital costs of $593/Kw, fixed O&M of $10/Kw/yr., variable O&M of 0.5 mills/Kwh, capacity factor of 85%, and a real capital charge rate of 10.4%.

Sources: Environmental Protection Agency, Electric Power Research Institute, CRS estimates.

As indicated, the analysis strongly suggests that natural gas pricing is the most important variable in determining generating costs from such plants. In the base-case analyses, fuel costs represent about two-thirds of the total costs (including capital charges). In the high cost analyses, fuel costs represent about three-fourths of the total costs. The importance of fuel costs is lessened a little by continuing improvements in generating efficiency. However, it is likely to remain the dominant cost factor over the time period discussed here. This variable would also have to be factored into any decision about existing coal-fired versus newly constructed natural gas-fired capacity.

Air Quality and Coal-fired Generation

If restructuring further encourages the increased utilization of existing coal-fired capacity in lieu of constructing new capacity and discourages energy conservation and development of cleaner technology because of low baseload pricing or other factors, the short- to mid-term effects could be increased air pollution. Operating an additional 23,000 Mw of coal-fired capacity would have significant air emissions, particularly for CO_2, and, depending on the fate of various EPA rulemakings, on NO_x. As SO_2 emissions are currently capped by title IV of the 1990 Clean Air Act Amendments, the effects of restructuring on SO_2 emissions should be negligible.

The most substantial effects of restructuring would be for carbon dioxide emissions, because they are currently uncontrolled. CO_2 emissions from 23,000 Mw of coal-fired capacity would be about 200,000,000 short tons, an increase of about 11% over 1996 CO_2 emissions by coal-fired electricity generation. This compares with emissions from a natural gas combined-cycle equivalent capacity of about 80,000,000 short tons, or a difference of 120,000,000 short tons.

Calculating the potential effects on NO_x emissions is more difficult as existing sources could be controlled under several provisions of the Clean Air Act (see table 3). For example, the final rule for the NO_x reduction program under section 407 of title IV of the Clean Air Act Amendments of 1990 was promulgated in 1996.[34] The rule will reduce the NO_x emission rate of coal-fired facilities examined here to an average of 0.48 lb/mmBtu. Based on this result, emissions from 23,000 Mw would come to about 480,000 tons, an increase of about 9% over 1996 coal-fired NO_x emissions. This would compare with emission of 70,000 tons from equivalent natural gas combined-cycle technology, or a difference of 410,000 tons.[35]

As identified earlier, other control possibilities, such as implementation of EPA's Ozone Transport Rule, also could reduce these emissions substantially.[36] Under this regulation, NO_x

[34] Environmental Protection Agency, "Acid Rain Program; Nitrogen Oxide Emission Reduction Program," 61 *Federal Register* 67111-67264 (December 19, 1996).

[35] Based on an average BACT determination of 0.1 lb./mmBtu. See Larry Parker, *Nitrogen Oxides and Electric Utilities: Revising the NSPS*. CRS Report 96-737, updated October 13, 1998.

[36] For a discussion of the transport rule, see Larry Parker and John Blodgett, *Air Quality: EPA's Ozone Transport Rule, OTAG, and Section 126 Petitions — A Hazy Situation?* CRS Report 98-236, updated July 14, 2000.

emissions across a 21-state area are "capped" at a specific level beginning September 30, 2007. That level of emissions can not be exceeded regardless of the electric utility industry's structure.

Similarly, successful prosecution of the NSR enforcement actions could impose additional control requirements on existing coal-fired facilities.[37] Under a restructured electric generating market, increased pollution control requirements would adversely affect the economics of affected facilities, which would become more expensive to operate. Increased capital and operating costs would make coal-fired capacity less attractive in a more competitive system.

As discussed above, existing coal-fired facilities are particularly vulnerable to future regulation of several pollutants. To illustrate the sensitivity of these facilities to increased pollution-control costs, CRS analyzed a representative sample of such potential costs. The results are presented in table 5. As indicated, control costs for each of these pollutants would add about 10% or more to the total generation costs from existing coal-fired facilities. Combinations of control measures would raise these costs even more.[38] With new natural gas combined-cycle technology potentially available for 2.5 cents/Kwh, increased air pollution control represents a real threat to the continuing operation of at least some existing coal-fired capacity.

These environmental concerns are not necessarily hypothetical. For example, member states of the Ozone Transport Commission (OTC) have agreed to stringent nitrogen oxide controls on stationary sources, including electric generating plants, in ten northeastern states. Depending on how the states and utilities choose to implement the program, selective catalytic reduction (SCR) or other control devices may have to be installed on some coal-fired power plants. This could also be the result of EPA's Ozone Transport Rule and/or a successful Section 126 petition with respect to interstate ozone pollution.

For natural gas combined-cycle facilities, the major *potential* environmentally related cost increase would be control of carbon dioxide.[39] *If* a new natural gas combined-cycle plant were required to offset all its potential CO_2 emissions under a future emissions cap, it could increase operating costs by about 0.2 cents/Kwh.[40] This would raise the total production costs for such facilities to 2.6-2.7 cents/Kwh, or 3.6-3.7 cents/Kwh if the high-cost gas scenarios were operative.

Table 5. Potential Pollution Control Cost for Existing Coal-fired Power Plants (500 Mw, 1995$)

Cost Factor	Nitrogen Oxides	Carbon Dioxide	Mercury	Sulfur Dioxide
Capital Costs	$49/Kw	0	$40.5/Kw	$190/Kw

[37] For an update on events surrounding EPA NSR enforcement activities, see: Larry B. Parker and John E. Blodgett, *Air Quality and Electricity: Initiatives to Increase Pollution Control*, CRS Report RS20553, updated December 28, 2000.

[38] Readers are cautioned not to simply add the incremental costs of these control measures together. There may be overlaps or efficiencies to be gained from controlling some of pollutants together that are presented in Table 5.

[39] The costs estimates cited above already include installation and operation of SCR. Natural gas plants emit very minor amounts of sulfur dioxide and mercury.

[40] This cost is very speculative. For a further discussion, see Larry Parker, *Coal Market Effect of CO2 Control Strategies as Embodied in H.R. 1086 and H.R. 2663*, CRS Report 91-883, December 13, 1991.

Fixed O&M	$4.27/Kw	0	$6.9/Kw	$6.8/Kw
Variable O&M	0.023 cents/Kwh	0.2 cents/Kwh	0.04 cents/Kwh	0.1 cents/Kwh
Total Costs of Control	0.17 cents/Kwh	0.2 cents/Kwh	0.22 cents/Kwh	0.53 cents/Kwh
Total Production Costs	2.17 cents/Kwh	2.2 cents/Kwh	2.22 cents/Kwh	2.53 cents/Kwh

Control Assumptions: For nitrogen oxides — installation of Selective Catalytic Reduction (SCR) with 70% removal; for carbon dioxide — buying carbon offsets for 50% of emissions at $5 a ton; for mercury — installation of carbon injection with spray cooling and fabric filter; for sulfur dioxide — installation of flue-gas desulfurization (FGD) with 95% removal.

Sources: U.S. EPA, Office of Air and Radiation, *Analyzing Electric Power Generation under the CAAA* (July 1996); and Larry Parker, *Coal Market Effects of CO_2 Control Strategies as Embodied in H.R. 1086 and H.R. 2663*, CRS Report 91-883, December 13, 1991.

ASSESSING THE IMPACTS OF RESTRUCTURING

Emissions from electricity generation are determined by an interactive process involving a utility industry and an environmental regulatory system that are both undergoing change. The dynamic linkages between electricity generation, resulting emissions, and pollution control make it difficult to separate out one factor (in this case, electricity restructuring) for analysis. The difficulty in doing this has been illustrated by various studies attempting to estimate the impact of restructuring on the environment.

For example, an early component of electricity restructuring is the Federal Energy Regulatory Commission's (FERC) Order 888 that promotes wholesale competition through open, non-discretionary access to transmission services to all participants in the wholesale generation market. In developing the Order, FERC conducted an environmental impact statement (EIS) to examine the implications of the proposed Order for emissions of pollutants by affected generating facilities. This assessment covered only a limited part of what would be affected by a comprehensive restructuring of the electricity generating industry; specifically, the Order is limited to the transmission of wholesale electricity, about 10% of total sales. Nevertheless, studies of the rule, including those critical of FERC's analysis, illustrate the difficulties in isolating the impacts of restructuring from other factors present in the system.

FERC issued its findings in a draft EIS[41] in November 1995. From two baselines — projections about electricity generation without the proposed rule — FERC analyzed changes in electricity generation that might result from the proposed rule, and the consequent changes in emissions that would therefore be expected. The two baselines differed in assumptions about the relative prices of gas and coal. Based on the models used by FERC and the assumptions adopted, the analyses indicated that the proposed rule would have a small effect on emissions. In general, through 2010, assumptions that favor gas could slightly decrease overall emissions, and assumptions that favor coal could slightly increase overall emissions. A regional analysis similarly found relatively small effects. Given the modest environmental impacts, FERC concluded that there was no need to undertake mitigation — although it

---

[41] Federal Energy Regulatory Commission, *Promoting Wholesale Competition through Open Access Non-Discriminatory Transmission Services by Public Utilities (RM95-8-000)* ... Draft Environmental Impact Statement (November 1995) FERC/EIS-0096D.

discussed options — and in fact concluded that it had little appropriate authority to require any mitigation.

Comments on the draft were numerous; they are summarized in the final EIS issued in April 1996.[42] Three issues received particular attention. Two sets of comments addressed two aspects of the analyses that commenters argued could have underestimated potential increases in emissions. A third set of comments focused on the issue of mitigation.

One set of these comments concerned the possibility that restrictions built into FERC's analysis on the amount of power that could be transmitted among regions unduly limited projections of the amount of electricity generated and exported from high-emitting, coal-fired sources in the Ohio River valley. These comments[43] suggested that the rule would increase the amount of power transported, leading to additional construction of more transmission capacity if necessary, and would thus result in more emissions than projected. In particular, it would increase NO_X emissions that could be expected to affect the Northeast. As a result, for its final EIA, FERC added further analysis of this possibility, but concluded the effects would not be significant.[44]

Another set of comments argued that the rule would have the effect of decreasing electricity prices and therefore would likely increase demand, leading to the generation of more electricity than assumed in the base cases. FERC basically said this possibility would be a second-order effect that lay outside appropriate analysis.[45] Despite FERC's response, ignoring demand seems unrealistic. As noted earlier, electricity demand is a critical component in assessing emission-related impacts. However, the model FERC used for its analysis is incapable of analyzing the price-demand effects of restructuring because its demand assumption is exogenous to the model. FERC chose to assume that the lower prices of restructuring would not result in any increase in electricity demand from baseline conditions — an unlikely outcome. To ignore the price-demand relationship reduces confidence in FERC's conclusion.

The third set of comments, on mitigation, ranged from those supporting FERC's conclusion that there was nothing that needs mitigating to arguments that FERC was obligated and has the authority to require mitigation.[46]

Subsequent reports challenge the FERC analysis. In April 1997, the Natural Resources Defense Council, Public Service Electric and Gas Co., and Pace University's Mid-Atlantic Energy Project jointly issued a report evaluating the contribution of utility generating companies to air pollution. Presenting data indicating that "the 'lowest cost' producers of electricity" are often "some of the highest emitters of pollutants," the authors concluded that

> In order to implement fair competition and to prevent a considerable increase in electric utility emissions due to increased use of older, higher-emitting units, the restructuring process should apply consistent environmental standards to all competitors.[47]

---

[42] Federal Energy Regulatory Commission, *Promoting Wholesale Competition through Open Access Non-Discriminatory Transmission Services by Public Utilities (RM95-8-000) ...* Final Environmental Impact Statement (April 1996) FERC/EIS--0096, Appendix J.

[43] See, for example, Alliance for Affordable Energy, et al., *Joint Comment on Draft Environmental Impact Statement* (February 1, 1996), p. 32.

[44] FERC, Final EIS, pp. J-34 - J-39 and pp. 6-25 - 6-41.

[45] See FERC, Final EIS, pp. J-69 - J-70.

[46] Mitigation is discussed in Chapter 7 of the EIS; comments are discussed in the Final EIS at pp. J-78 - J-105.

[47] Natural Resources Defense Council, et al., *Benchmarking Air Emissions of Electric Utility Generators in the Eastern United States,* 2nd Edition (April 1997), p. 41.

In January 1998, the Northeast States for Coordinated Air Use Management (NESCAUM) issued a report concluding that recent trends contradict FERC's finding in the EIS. Specifically, NESCAUM presented data challenging two assumptions that had led to the EIS conclusion that emissions growth would be negligible. Contrary to the EIS analysis, NESCAUM shows that between 1995 and 1996, coal-fired generation increased while natural gas-fired generation declined and that growth in the use of interregional power transmission had "outstripped FERC's longer-term growth assumptions."[48] According to NESCAUM, these

> preliminary findings suggest that increased competition is contributing to increased emissions at coal-intensive utilities, and that some form of mid-course public policy correction may be necessary. These findings underscore the need for comprehensive efforts to document the impacts of restructuring on air quality, and lend impetus to state and federal efforts to establish adequate emissions tracking and disclosure systems. Moreover, these findings suggest that equitable environmental standards must be made an integral part of ongoing competitive reforms.[49]

Thus, environmental interest groups continue to warn that FERC underestimated emissions resulting from its rule — and that restructuring portends even greater impacts; and that, therefore, mitigation of the effects of the rule and of restructuring is necessary. However, as suggested above, increased emissions may be the result of existing trends in the industry, and not strictly due to restructuring. As noted, renovating coal-fired capacity has been an increasing trend in the industry for over a decade. As the NESCAUM data reflect a time period before implementation of Order 888, ascribing emission increases solely to restructuring is debatable. This situation illustrates the difficulties in assigning cause to potential emission increases over the next 5-10 years from existing coal-fired facilities.

CONCLUSION

The relationship between restructuring electricity generation and environmental consequences is not a simple one. The environmental outcome will result from an interactive, iterative process of many changes in existing trends affecting electricity generation. The two most crucial trends are: (1) decisions with respect to meeting future electricity demand, including the renovation of existing generating capacity, choice of new generating technologies for new construction, and enhancement of transmission capacity; and, (2) decisions with respect to implementing existing environmental regulations, and the potential approval of future environmental regulations. Restructuring would influence each of these trends to varying degrees, encouraging some, such as renovating existing capacity, and challenging others, such as existing environmental regulations.

[48] Northeast States for Coordinated Air Use Management, *Air Pollution Impacts of Increased Deregulation in the Electric Power Industry: An Initial Analysis* (January 15, 1998), p. 1. At [http://www.nescaum.org/about.html]

[49] Ibid., p. 2.

Electricity Demand

Restructuring and the other trends underway point to changes in demand and technological developments that will ultimately be reflected in environmental consequences. To the extent restructuring and the other changes lead to a more efficient generation industry, baseload prices should decline, which would be expected to lead to higher demand and greater consumption. Lower baseload prices could encourage owners of existing coal-fired facilities with low operating costs to extend and enhance electricity generation from such facilities rather than risk investing in new construction. CRS estimates that between 23,000 and 53,000 Mw of existing coal-fired capacity is currently underutilized and could be made available if economics and transmission capacity justified such a decision. Renovating existing coal-fired capacity has been an increasing trend in the industry for over a decade. The more competitive generating market of a restructured electric utility industry could further encourage this trend.

Reduced baseload electricity prices also change the signals affecting consumer choices related to energy efficiency. Lower baseload electricity prices could diminish the incentive to invest in increased conservation, such as more efficient refrigerators or insulation. Cost-considerations may also work against power generation by renewables such as solar, wind power, and geothermal, which currently are not cost-competitive with natural gas or coal technologies. It may also work against nuclear, which is a capital intensive technology, and which has contradictory environmental implications — being essentially free of air emissions, but posing waste disposal problems that some see as more hazardous and less controllable. Finally, cost concerns may further encourage natural gas-generated power in new construction (the existing technology of choice), which is more environmentally friendly than coal or oil alternatives.

But at the same time, if prices reflect marginal costs, the price signal is likely to dampen peak demand, which typically is met by the most costly and inefficient generating capacity — thereby leveling the demand curve. Higher prices for peak load power could strengthen the signal for load management — conservation measures that reduce peak usage, such as automatic shutoffs of hot water heating during peak demand. Also, to the degree consumers are given the choice of electricity suppliers, they may create new markets for different types of generation by basing their decisions on factors other than economics, such as environmental ones. One such possibility is "green pricing," where some consumers choose to purchase electricity that costs more economically but costs less environmentally — such as that produced by renewables. Such a "green market" is being developed in California, but it is too early to anticipate the size that it may achieve.

How these differing effects play out will determine the potential for increased emissions from restructuring. As indicated, some, such as renovating existing coal-fired capacity, represent a furthering of an existing trend. Others, such as green pricing, represent a new trend created by restructuring. Although the overall effect on emissions is difficult to assess, involving several currently unquantifiable variables, the most substantial environmental effect in the short- to mid-term is likely to come from enhanced operation of existing coal-fired capacity. Whether one can ascribe that effect to general trends in the industry or to restructuring is debatable.

Air Quality Regulations

Restructuring, combined with the outcome of the other trends, has the potential to increase emissions of some pollutants of concern; the question is whether existing (or proposed) regulatory limits on those emissions would effectively prevent adverse effects.

- For SO_2, restructuring is unlikely to have any effects on emissions. The CAA requirements statutorily "cap" the nation's utility SO_2 emissions, making industry structure essentially irrelevant. Increasing numbers of participants may make monitoring and enforcement more demanding, but the SO_2 program contains substantial penalties for non-compliance, and no compliance difficulties have emerged to date.
- For NO_x, the potential of extended and enhanced coal-fired capacity utilization encouraged by restructuring could significantly increase emissions. NO_x emissions from an additional 23,000 MW of coal-fired capacity could be in the range of 480,000 tons, compared with about 70,000 tons if that electricity was generated from natural gas. However, several EPA regulatory actions could reduce or eliminate that potential increase. For example, EPA's Ozone Transport Rule would in effect set a "cap" on emissions in 21 eastern states where they currently contribute to unacceptable ozone pollution; these 21 states are where the majority of potential coal-fired related NO_x emission increases could occur. However, the primary implementation of the process lies with the states. As a result, the NO_x control program would be administratively more complicated and could be less economically efficient than the SO_2 control program. A system of state-based programs to control NO_x emissions might dovetail with the current electricity generation system in which state regulation plays a large role; but if restructuring leads to a more regionally-based, competitive electricity generating system, then implementing a NO_x control program based on state programs could lead to industry segments being subject to inconsistent requirements in the various states. Indeed, the inconsistencies could constitute barriers — or opportunities — to the restructuring process. But if the process works, then there would be no increase in NO_x emissions in the 21-state region, and the structure of the industry would be irrelevant. If this process is delayed, other regimens, including section 126 petitions, are available for relief.
- For CO_2, the potential of extended and enhanced coal-fired capacity utilization encouraged by restructuring could significantly increase emissions. CO_2 emissions from an additional 23,000 Mw of coal-fired capacity would be about 200,000,000 short tons, compared with 80,000,000 tons if that electricity was generated from natural gas. CO_2 emissions are not controlled by the Clean Air Act, nor does there appear to be any readily applicable provision that could used to control such emissions. *If* the U.S. ratifies the Kyoto Agreement, it would effectively "cap" emissions, and restructuring would become irrelevant in terms of emission increases. But how emissions would be controlled and how reductions would be allocated and implemented would remain to be determined by future domestic legislation.
- For Hg, increased utilization of coal-fired capacity would result in increased Hg emissions, although uncertainty exists as to how much that increase would be. Studies have been completed that could be the basis for regulation under existing

CAA authorities, if EPA were to conclude controls are necessary.[50] How well these would mesh with a more competitive electric generating market is unclear.

Ultimately, whether future developments in electricity generation will lead to pollution increases of health and environmental concern depends on the effectiveness of CAA requirements and of EPA implementation and enforecement (or on future enactments of new controls) — and, in the case of CO_2, on whether the U.S. ratifies the Kyoto Convention and its requirements are implemented.

Responses

If one has confidence that these authorities will prove adequate to protect human health and the environment and will be effectively implemented, one may be comfortable in adopting the stance that "no action" is necessary to address any emissions implications of restructuring proposals. The result might not be as cost-effective as a regulatory regime more tied to a competitive market (such as a SO_2 style "cap and trade" program), but except for CO_2 (for which the need for controls is highly contentious), the CAA provides control authorities — notably for emissions of particular concern, NO_x and Hg.[51]

Conversely, if one fears that existing approaches to pollution control will not be effective or not be implemented, or that controls on CO_2 are requisite, then one is likely to press actions for immediate response. Those having such concerns could be expected to pursue several actions to address perceived unacceptable environmental impacts, such as:

- aggressively using the existing regulatory regimen to address environmental deterioration;
- proposing to embed environmental protections in any restructuring programs; and
- proposing revisions of appropriate environmental statutes, in particular, the CAA — and supporting prompt ratification of the Kyoto Agreement and enactment of appropriate implementation legislation.

Those seeking to assure rigorous application of existing air pollution controls could aggressively use the citizen suit or section 126 provisions of the CAA to press their case whenever they perceive inadequate or ineffective implementation. While this may be effective for controlling NO_x and Hg emissions, it would not appear to be a fruitful course with respect to CO_2 emissions for which any controls are only prospective.

The case for adding environmental protections to restructuring proposals depends in part on the extent to which restructuring itself is a likely cause of more pollution; or if the restructured industry were to pose implementation, enforcement, or other problems that the current regulatory structure proves ill-equipped to cope with. (For some, attaching environmental issues to restructuring may be a surrogate for debates on a comprehensive review of the CAA or for CO_2 controls.) As indicated by the analysis of this paper,

[50] Specifically, see: Environmental Protection Agency, *Mercury Study Report to Congress*, EPA-452/R-97-003 (December 1997).

[51] While Hg emissions from electric utilities are not currently regulated, the 1990 CAA Amendments provide EPA with the authority to do so based on studies mandated by the Act (section 112(n)).

restructuring's role is not clear; it is quite possible that some utility emissions of concern could increase as a result of other trends underway, and in fact that may be happening. As the NESCAUM report indicates, coal-fired generation appears already to be on the rise, before restructuring efforts such as FERC's rule to promote wholesale competition through open access to transmission services could be having an effect. Likewise, monitoring shows that (unregulated) CO_2 emissions have risen since 1990.[52] Thus, tying environmental protection to restructuring might fail to address actual causes for increased emissions.

Responding to any problems that arise by then revising environmental statutes may seem risky to those who already perceive significant environmental risks from changes in electricity generation. They could point to the history of acid rain legislation as an example of the risk: it took about 10 years from the time legislation was first proposed to address acid rain to enactment of a program — which required movement of a comprehensive set of CAA amendments. The lesson many draw is that consideration and enactment of environmental legislation separate from legislation that might have the potential to cause environmental problems can be delayed and difficult. From this perspective, at least, it could make sense that if Congress is to enact restructuring legislation, to attach to those bills any potentially necessary environmental responses even if the problem is not solely due to restructuring.[53]

All in all, as the preceding analysis and discussion indicates, the potential for environmental deterioration from restructuring electricity generation is difficult to project — both because various technical and economic changes are affecting the industry at the same time and because of an evolving policy context. As a result of this uncertainty, those who are focused on preventing environmental deterioration tend to take a precautionary stance, to propose immediate preventative measures, and to argue that such measures be attached to available legislative vehicles. In contrast, those who doubt that there will be significant environmental effects or who are focused on the substantial regulatory structure in place tend to take a wait-and-see position. Further complicating this picture is that some attitudes about restructuring are related to and partly a surrogate for a more fundamental debate that is underway because of global climate change concerns — about the future direction of energy use in the United States and the federal role in affecting it.

[52] Larry Parker and John Blodgett, *Global Climate Change: Reducing Greenhouse Gases — How Much from What Baseline?* CRS Report 98-235, March 11, 1998.

[53] For information on legislative proposals relating restructuring to environmental controls, see Larry Parker and Amy Abel, *Electricity: The Road Toward Restructuring*, CRS Issue Brief IB10006 (updated regularly); and Larry Parker, *Electricity Restructuring: Comparison of Comprehensive Bills,* CRS Report RL30087, (updated regularly); and, Larry Parker, *Electricity Restructuring and Air Quality: Comparison of Proposed Legislation*, CRS Report RS20326 (July 26, 2000).

Chapter 5

THE SUPREME COURT UPHOLDS EPA STANDARD-SETTING UNDER THE CLEAN AIR ACT: *WHITMAN V. AMERICAN TRUCKING ASS'NS*[*]

Robert Meltz and James E. McCarthy

INTRODUCTION

On February 27, 2001, the Supreme Court handed down its eagerly awaited decision in *Whitman v. American Trucking Associations.*[1] The case stems from EPA's controversial promulgation in 1997 of revised national ambient air quality standards (NAAQSs) for ozone and particulates, under the Clean Air Act (CAA). More fundamentally, the legal theories in the case raise basic questions about standard-setting under the Clean Air Act and the constitutional limits on congressional delegation to agencies of standard-setting authority.

BACKGROUND

NAAQSs lie at the very heart of the Clean Air Act. These standards prescribe maximum pollutant concentrations for ground-level, outdoor air, and have been promulgated by EPA for six pollutants,[2] including ozone and particulates. The NAAQSs determine the stringency of emission limits that each state must impose on individual stationary sources of the six pollutants, to achieve the NAAQSs within its borders. NAAQSs come in two forms: "primary NAAQSs" protect the public health, while "secondary NAAQSs" protect the public "welfare" (non-public health effects).[3] Once a NAAQS has been promulgated, EPA must review it (and

---

[*] Excerpted from CRS Report RS20860. March 28, 2001.

[1] 121 S. Ct. 903 (2001). The decision adjudicated both EPA's petition (No 99-1257) from the decision below, and an industry cross-petition (No. 99-1426).

[2] 40 C.F.R. §§ 50.4-50.12.

[3] CAA § 109(b); 42 U.S.C. § 7409(b).

the "criteria documents" on which it is based) every 5 years, and make such revisions "as may be appropriate."[4]

In 1997, EPA revised the NAAQSs for ozone and particulate matter, making them stricter. Given the perceived impact of these more stringent standards on the economy, it was unsurprising that numerous legal challenges were brought – with two members of Congress (Rep. Bliley and Sen. Hatch) filing as amici on the side of the challengers. Pursuant to CAA requirement, the suit was filed in the U.S. Court of Appeals for the District of Columbia Circuit.

In May, 1999, the D.C. Circuit ruled 2-1 that various deficiencies in EPA's promulgation of the two NAAQSs required that they be sent back to EPA for further consideration.[5] Among other things, the two-judge majority held that EPA's reading of the CAA section governing the setting of primary NAAQSs gave the agency too much discretion, and thus violated the constitutional "nondelegation doctrine." It also rejected industry's position that EPA, in arriving at primary NAAQSs, may consider the costs of implementation. Finally, it ruled that EPA could not enforce its revised primary NAAQS for ozone, owing to its being an eight-hour standard, rather than the one-hour standard envisioned by CAA nonattainment-area provisions added to the Act in 1990 (see further discussion on page 4). Five months later, the three-judge panel made minor modifications in its opinion, but the full court refused to grant rehearing *en banc* (all the judges of the court sitting).[6] In May, 2000, the Supreme Court took the case.[7]

What the Supreme Court Said

The Supreme Court in *American Trucking* gave EPA a unanimous victory on the two major issues in the case: consideration of costs, and nondelegation doctrine. Justice Scalia authored the opinion of the Court, with various justices writing separate concurrences to note differences as to rationale, but not as to holding.

Consideration of Costs

The Court affirmed the D.C. Circuit decision (which, in turn, had endorsed existing case law of the circuit) in holding that when promulgating primary NAAQSs, or *revised* primary NAAQSs, EPA may not consider the costs of implementing the new standard. Health impacts, and health impacts alone, are to be the touchstone. The governing standard in the statute, the Court said, made this clear: section 109(b)(1) instructs EPA to set primary NAAQSs "the attainment and maintenance of which ... are requisite to protect the public health" with an "adequate margin of safety."[8]

Industry's arguments that considerations other than the health impacts of pollutants were cognizable could not overcome the directness of the above statutory text. For example, industry contended that a very stringent NAAQS might close down whole industries, thereby

---

[4] CAA § 109(d)(1); 42 U.S.C. § 7409(d)(1).

[5] 175 F.3d 1027 (D.C. Cir. 1999).

[6] 195 F.3d 4 (D.C. Cir. 1999).

[7] 120 S. Ct. 2003 (petition of EPA, granted May 23, 2000); 120 S. Ct. 2193 (petition of American Trucking Ass'ns, granted May 30, 2000).

[8] 42 U.S.C. § 7409(b)(1).

impoverishing the workers dependent on that industry and, in turn, reducing their health. A health-based standard such as the primary NAAQS should include these indirect impacts, industry asserted. The Court, however, pointed to numerous other CAA sections where Congress had explicitly allowed consideration of economic factors, concluding that had it intended to allow such factors under section 109(b)(1), it would have been more forthright – particularly given the centrality of the NAAQS concept to the CAA's regulatory scheme. Looking for such a forthright "textual commitment" of authority for EPA to consider costs, the Court found none. Its conclusion: section 109(b)(1) "unambiguously bars cost considerations from the NAAQS-setting process."

Nondelegation Doctrine

The most controversial portion of the D.C. Circuit's majority opinion was its embrace of a long-moribund constitutional principle known as the "nondelegation doctrine." This separation-of-powers doctrine derives from Article I of the Constitution, which vests "[a]ll legislative Powers" in Congress. Not surprisingly, the Supreme Court reads this vesting provision loosely, recognizing that Congress routinely delegates quasi-legislative powers to non-Article I bodies. In particular, Congress frequently commits to the specialized expertise of executive-branch agencies the task of rulemaking in technical areas — such as air pollution control. The nondelegation doctrine says that such delegations pass constitutional muster only if Congress gives the agency an *intelligible principle* to guide its exercise of that authority.

The majority opinion below found that EPA had construed CAA section 109 so loosely as to render it an unconstitutional delegation. The court agreed with the *factors* used by the agency to assess the public health threat posed by air pollutants. But, it said, EPA had articulated no intelligible principle for translating the factors into a particular NAAQS, nor is one apparent from the statute. Given that both ozone and particulates are non-threshold pollutants (adverse health effects occur at any concentration above zero), *some* public health threat has to be tolerated if EPA is to avoid shutting down entire industries. The agency, in the court's view, had articulated no standard for determining *how much*.

In invoking the nondelegation doctrine, the D.C. Circuit drew considerable attention. It was the first time in 65 years that the nondelegation doctrine had been successfully used, and raised serious implications for how Congress delegates standard-setting authority to agencies generally. Commentators pointed to other federal statutes – such as the Corps of Engineers wetlands permitting program under the Clean Water Act, and the rulemaking authority conferred by the Occupational Safety and Health Act – as vulnerable to nondelegation-doctrine challenge, should the D.C. Circuit be affirmed on appeal.

The Supreme Court, however, reversed. The scope of discretion allowed by section 109(b)(1), the Court said, is "well within the outer limits of our nondelegation precedents." Under section 109(b)(1), primary NAAQSs are to set at levels "requisite" to protect public health – "requisite" being argued by the United States, and accepted by the Court, as meaning "sufficient, but not more than necessary." To be sure, acknowledged the Court, more guidance must be furnished the agency when the agency action is to have broad scope – as here, where the revised NAAQSs affect the entire U.S. economy. But even for sweeping regulatory schemes, the Court disclaimed any demand that statutes provide a "determinate criterion" for saying precisely how much of the regulated harm is too much. EPA may therefore be allowed discretion to determine how much of a public health threat from ozone and particulates (recall, they are non-threshold pollutants) it will tolerate at non-zero levels.[9]

Issues Involving Implementation of the Revised Ozone NAAQS

EPA lost, again unanimously, on two issues arising from its policy for implementing the revised ozone NAAQS in non-attainment areas. First, the Court rejected EPA's argument that the policy did not constitute final agency action ripe for review.

The Court then proceeded to the second issue: which CAA provisions govern the ozone nonattainment-area implementation policy. This calls for some background. The CAA imposes restrictions on nonattainment areas over and above those that the Act imposes generally. These additional nonattainment-area restrictions are found in Title I, Part D of the statute. Subpart 1 of Title D contains general nonattainment regulations that apply to every pollutant for which a NAAQS exists. Subpart 2 of Part D addresses ozone in particular. The dispute before the Court was whether Subpart 1 alone, or rather Subpart 2 or some combination of Subparts 1 and 2, controls the implementation of the revised ozone NAAQS in nonattainment areas. EPA, in its implementation policy, took the former, Subpart-1-only course. The problem it faced was that Subpart 2 contemplated a 1-hour ozone NAAQS, reflecting the ozone standard existing when Subpart 2 was enacted in 1990. The revised ozone NAAQS, however, embodied an 8-hour standard. Thus, some Subpart 2 provisions – in particular, the nonattainment-area classification scheme that identified the requirements to be imposed depending on an area's degree of nonattainment – did not fit the new ozone NAAQS.

The Court found that EPA could not ignore Subpart 2 entirely, as it had done. Whatever awkwardness of fit results from applying Subpart 2 to the new ozone standard, it cannot, said the Court, justify "render[ing] Subpart 2's carefully designed restrictions on EPA discretion utterly nugatory once a new standard has been promulgated ...." One example of the discretion-limiting nature of Subpart 2: under Subpart 1, EPA may extend attainment dates for as long as 12 years; under Subpart 2, only 2 years (though Subpart 2's attainment deadlines stretch from 3 to 20 years depending on the severity of an area's ozone pollution). The Court left it to EPA "to develop a reasonable interpretation of the [CAA's] nonattainment implementation provisions" for the revised ozone NAAQS.

LEGAL COMMENTARY

As to the nondelegation issue in *American Trucking*, EPA had gone to the Supreme Court with the stronger arguments. It was not a foregone conclusion, however, that EPA would win unanimously, as it did, even though the Court had on many occasions sustained federal statutes containing standards as loose as, or looser than, that in the CAA governing NAAQS setting. A few justices voting for at least some resuscitation of the nondelegation doctrine was widely deemed a possibility on the ground that the Court in other areas recently has revealed an interest in cabining congressional power and discretion – and not only when the federal-state balance is implicated. Though it has been 66 years since the last successful

[9] The Supreme Court also soundly rejected the D.C. Circuit's view that an unconstitutionally standardless delegation of power can be remedied by the agency's adoption of an interpretation confining its own discretion. *Congress*, it said, must lay down the intelligible principle.

nondelegation-doctrine challenge, the Court has not hesitated to reverse longstanding patterns in its constitutional jurisprudence when doing so furthered the agenda of a contingent of the justices. Very likely, the Court's refusal to bring back the nondelegation doctrine stemmed in part from the Court's view that "in our increasingly complex society, replete with ever changing and more technical problems, Congress simply cannot do its job absent an ability to delegate power under broad general directives."[10] Moreover, the task of rewriting federal statutes necessitated by a reinvigorated nondelegation doctrine might be one of daunting magnitude. From the environmental area to antitrust to civil rights, federal laws abound that give agencies only the broadest of guidance.

Likewise, EPA had the better argument when it came to inclusion of costs in the setting of NAAQSs. As a critic of using legislative history to divine the meaning of statutes, Justice Scalia restricted his analysis for the Court to examination of the statutory text itself. But the concurrence by Justice Breyer reveals a legislative history from the enactment of the 1970 CAA that sides unambiguously with EPA's keep-costs-out position.

For the foregoing issues then, the *American Trucking* decision largely restores the status quo ante. As before the filing of this case, a lax jurisprudence under the nondelegation doctrine and the impermissibility of considering costs in setting NAAQSs are once again regarded as relatively settled law. Some commentators have noted the Court's refusal in *American Trucking* to defer to EPA's interpretation of the CAA on ozone standard implementation, and speculated that the legacy of the decision may lie in its signalling a desire by the Court to lessen the degree of judicial deference to agency decisionmaking. But it is premature as yet to draw this conclusion.

REMAINING ISSUES

The Court's decision in *Whitman v. American Trucking* is not the end of court proceedings in the case. Numerous issues remain before the D.C. Circuit, and EPA's next attempt at an implementation plan may also be subject to court challenge.

Developing an implementation plan that embodies a "reasonable interpretation" of the Act's nonattainment implementation provisions for the revised ozone NAAQS, as the Court mandated, is not an easy task. EPA has said that an 8-hour standard of 0.09 ppm would have "generally represent[ed] the continuation of the [old] level of protection,"[11] but the new standard is, in fact, set at a more stringent level of 0.08 ppm. Thus, the statute's classification system contains a gap that does not address areas with readings of 0.08 - 0.09 ppm.

A second problem relates to the setting of attainment dates. In the 1990 amendments, Congress was specific in setting dates of attainment that ranged from 3-20 years from the date of enactment, depending on the severity of an area's ozone pollution. The Court read this specificity as denying EPA its previous broad discretion in setting attainment dates. But three of the Act's deadlines (for Marginal, Moderate, and Serious ozone nonattainment areas) have already passed. How EPA is to respond to this is unclear. The Court itself noted that the Act's

[10] Mistretta v. United States, 488 U.S. 361, 372 (1989).

[11] National Ambient Air Quality Standards for Ozone; Final Rule, 62 Fed. Reg. 38,856, 38,858 (1997), quoted in *Whitman* at 121 S. Ct. 903, 918.

method for calculating attainment dates "seems to make no sense for areas that are first classified under a new standard after November 15, 1990."[12]

In these circumstances, it would seem likely that whatever approach EPA may take will be subject to challenge by parties opposed to the new standards, with the potential for several additional years of litigation before the issues are resolved. Whether Congress should intervene to settle these matters is a possibility that few have discussed. While logical in many respects, such a legislative clarification would open a number of issues regarding the level of the new standards, the implementation measures to be required, and the nature of EPA's standard-setting authority that interested parties may not wish to have legislated.

[12] 121 S. Ct. 903, 918.

Chapter 6

AIR QUALITY AND ELECTRICITY: ENFORCING NEW SOURCE REVIEW*

Larry B. Parker and John E. Blodgett

INTRODUCTION

On November 3, 1999, the Justice Department filed seven lawsuits against electric utilities in the Midwest and South charging them with violations of the New Source Review (NSR) requirements of the Clean Air Act (CAA).[1] The seven utilities affected are American Electric Power (AEP), Cinergy, First Energy, Illinois Power, Southern Indiana Gas & Electric Company, Southern Company, Tampa Electric Company, or their subsidiaries. Filed on behalf of EPA, the lawsuits represent one of the largest CAA enforcement actions ever taken by EPA. Additional lawsuits have been filed by the State of New York and Connecticut (jointly), Vermont, and New Jersey against AEP,[2] and by a coalition of environmental groups also against AEP.[3] In addition, the Environmental Protection Agency (EPA) issued an administrative order against the Tennessee Valley Authority (TVA), alleging similar violations.[4]

Enacted as part of the 1977 CAA Amendments and modestly modified in the 1990 CAA Amendments, NSR is designed to ensure that newly constructed facilities, or substantially modified existing facilities, do not result in violation of applicable air quality standards. The NSR provisions establish permit requirements for constructing new major pollution sources

---

* Excerpted from CRS Report RL30432. January 31, 2000.

[1] The lawsuits also allege violation of the major modification provisions of section III, involving the New Source Performance Standards (NSPS). Copies of the lawsuits can be found at EPA's website: http://es.epa.gov/oeca/ore/aed/coal/mdex.html

[2] *New York and Connecticut v. American Electric Power Co.,* S.D. Ohio, No. C2-99-1182 (November 29,1999).

[3] The environmental groups involved are the Natural Resources Defense Council (NRDC), Citizens Action Coalition of Indiana, the Clean Air Council of Pennsylvania, the Hoosier Environmental Council, the lzaak Walton League, the Ohio Citizen Action, the Ohio Public Interest Research Group, Ohio Valley Environmental Coalition, the Sierra Club, U.S. Public Interest Research Group, and the West Virginia Environmental Council.

[4] *In re TVA,* EPA, No. CAA-2000-04-0008 (November 3, 1999).

as well as for making major modifications to existing major pollution sources. The specific requirements dictated by NSR depend on where the facility is sited. In attainment areas - those meeting the National Ambient Air Quality Standard (NAAQS) for a pollutant - the governing requirements are the Prevention of Significant Deterioration (PSD) provisions of the CAA. In non-attainment areas — those not in compliance with a NAAQS for one or more pollutant — the governing requirements are covered by non-attainment provisions. Some facilities can be subject to a combination of both, if the area is in attainment for some criteria pollutants[5] but not others. Meeting these permitting requirements can be a long and complex process, depending on the specific project, the pollutants involved, and the specific state and federal regulatory authorities involved.[6] In 1996, EPA proposed changes to NSR to streamline it. However, the proposals have been subject to considerable controversy, and a final rule has not yet been issued[7] These complexities and controversies, particularly with respect to major modifications of existing sources, become manifest in the enforcement suits filed by the Justice Department for EPA, and in the responses to them.

The suits represent a continuing effort by EPA to reduce pollution from existing sources, particularly coal-fired electric generating facilities. The primary pollutant of concern has been nitrogen oxides (NO_x), a criteria air pollutant that is also a precursor to ozone (itself a criteria pollutant), acid rain, fine particulates, and regional haze; other pollutants of concern include sulfur dioxide (SO_2), a criteria air pollutant and precursor to acid rain, fine particulates, and regional haze, and particulate matter (also a criteria air pollutant). Efforts by EPA and other groups to reduce these emissions have intensified recently both as a result of new regulatory proposals — to increase the stringency of the ozone and particulate NAAQS[8] and to reduce haze[9] — and as a result of the restructuring of the utility industry which some fear may result in increased emissions from existing plants.[10]

EPA's recent enforcement action raise several questions with respect to the NSR and efforts to further control emissions from existing sources. Three of them are discussed in this report: Why is EPA taking this action? Why use the NSR provisions? How does the NSR interact with other EPA actions currently underway?

---

[5] Pollutants for which EPA has set NAAQS are often called "criteria pollutants" after the criteria documents EPA prepares for setting the standard. For background on NAAQS and the criteria air pollutants and how the CAA is structured to ensure attainment of clean air, see *Environmental Laws: Summaries of Statutes Administered by the Environmental Protection Agency,* CRS Report RL30022, pp. 7-13; and John E. Blodgett, Larry B. Parker, and James E. McCarthy, *Air Quality Standards: The Decisionmaking Process,* CRS Report 97-722 ENR (June 24,1998).

[6] Many of the activities under the CAA, including many requirements specifically involving NSR, either reside with or can be and have been delegated to states (which can include territories, Indian tribal governments, and the District of Columbia). In essentially all cases, EPA can act in lieu of states to which authorities have not been delegated, or whenever states fail to take required actions.

[7] 61 *Federal Register* 38249-38344 (July 23, 1996).

[8] See John E. Blodgett and Larry B. Parker, *Air Quality Standards: EPA's Final Ozone and* Particulate *Matter Standards,* CRS Report 97-721 ENR.

[9] See James E. McCarthy, John E. Blodgett, and Larry B. Parker, *Regional Haze: EPA's Proposal to improve risibility in National Parks and Wilderness Areas,* CRS Report 97- 1010 ENR.

[10] For more information on air quality and electricity restructuring, see: Larry Parker and John Blodgett, *Electricity Restructuring: The Implications for Air Quality,* CRS Report 98-615 ENR (July 16,1999).

WHY EPA'S ENFORCEMENT ACTION?

"The EPA believes that there should be a significantly higher rate of NSR permits authorizing the construction of new or expanded facilities than is currently taking place."[11]

The CAA requires a preconstruction review of, and a permit for, almost any major modification of an air polluting source or any major new source. Assuming that a state has an EPA-approved State Implementation Plan (SIP), which spells out the state's strategy for complying with NAAQS, regulatory approval to construct the new source or modify the existing source must come from the appropriate state agency. To receive this "Permit to Construct," the applicant must show that the proposed source or modification will not result in, or exacerbate, violation of a NAAQS, either locally or downwind. In addition, applicants must show that their proposal will not result in local or downwind exceedences of increments of increased air pollution allowed under Prevention of Significant Deterioration (PSD) regulations in areas complying with NAAQS. It is this preconstruction review process that is called New Source Review (NSR).[12]

The NSR process is triggered for any new source that potentially could emit 100 tons annually (or less in some areas[13]) of any criteria air pollutant, and by any modification that will cause a significant increase in annual emissions (regulatorily defined as 40 tons for SO, and N0x[14]). The specific NSR requirements for affected sources depend on whether the sources involved are subject to the PSD or the non-attainment provisions.[15] If covered by PSD, the source is required to install Best Available Control Technology (BACT), which is determined on a case-by-case basis, and which cannot be less stringent than the federally determined New Source Performance Standard (NSPS) for that pollutant. If covered by non-attainment provisions, the source is required to install Lowest Achievable Emission Rate (LAER) and obtain applicable offsets for that particular area.[16] Like BACT, LAER must not be less stringent than the federal NSPS.

Despite the breadth of coverage suggested by NSR, few preconstruction permits have been issued to coal-fired power plants over the program's history.[17] If this situation is examined from the perspective of new construction, the lack of permits is not too surprising. Current U.S. coal-fired electric generating capacity is about 300,000 megawatts (MW), and has remained steady at that level for the last ten years.[18] As indicated in **Table I,** additions to

---

[11] Steven A. Herman, Assistant Administrator for Enforcement and Compliance Assurance, EPA. Letter to Chairman Inhofe (March 26, 1999), p. 2.

[12] Some restrict the term "NSR" to the review process in a non-attainment area only; the review process in an attainment area being called "PSD pre-construction review". This paper will use the term to indicate both. In addition, new and modified sources must meet New Source Performance Standards (NSPS).

[13] Title I of the 1990 CAA Amendments reduced the threshold definition of a major source for areas of the country substantially out of compliance with ambient air quality standards for ozone, carbon monoxide, and particulate matter.

[14] 40 CFR 52.24

[15] It should be noted that a source can be affected by the PSD requirements for one pollutant, and by the non-attainment requirements for another pollutant.

[16] For details on these provisions and their requirements, see *Clean Air Act, Part C - Prevention of Significant Deterioration of Air Quality, sections 160-169;* and. *Part D-Plan Requirements for Nonattainment Areas, sections 171-178.*

[17] Letter to Chairman Inhofe, p. 2.

[18] Data represent net summer capacity. Energy Information Administration, *Annual Energy Review 1998,* DOE/EIA-0384(98) (July 1999), p. 219.

coal-fired capacity, while greater than retirements, have not been significant. Capacity that began operation between 1989-1998 constitutes about 3% of total current coal-fired capacity.

Table 1. Capacity Additions and Retirements: 1989-1998 (net summer capacity)

Year	Capacity Additions (MW)	Retirements (MW)
1989	1,967	379
1990	3,063	175
1991	792	377
1992	498	254
1993	0	104
1994	540	461
1995	1,036	2
1996	1,611	16
1997	0	293
Total 1989-1998	**9,507**	**2,063**

Source: Energy Information Administration, *Inventory of Power Plants in the United States,* various years.

At first glance, emissions trends for electric utilities appear to confirm the generally static situation with respect to coal-fired capacity. As shown in **Figure** I, emissions of NO_x from coal-fired electric generating facilities have remained constant over the last ten years. In contrast, 802 emission trends indicate a significant drop in emissions in 1995, the first year of the acid rain control program under title IV of the 1990 CAA Amendments. Under title IV, I II powerplants were required to reduce their emissions by about 30% from baseline levels specified in the title. In addition, many units were required to install Low-NO_x burners to reduce NO_x emissions.[19] According to EPA, the affected phase I units emitted about 10 million tons of SO, in 1990, a number reduced to 5.3 million tons in 1995. Likewise, the NO_x provisions reduced emissions from these plants in 1996 by about 0.4 million tons from their 1990 levels.[20]

However, comparing the reductions achieved under title IV with the trends indicated in figure I shows that some of the title IV reductions have been offset by increases in emissions from other plants: specifically, while title IV caused a 4.7 million ton SO, reduction by 1995, emissions data indicates that overall emissions only went down about 3.6 million tons. Likewise, while title IV led to a 0.4 million ton reduction in NO_x by 1996, figure I reveals an overall reduction of only about 0.1 million tons. This slippage is not surprising as phase I does not cover all coal-fired facilities, but it does imply that within the relatively flat trends for coal-fired capacity and for air emissions, there are some sources of increasing emissions.

The dynamism in coal-fired generation is the continuing operation of existing coal-fired facilities. As indicated by **Figure** 2, despite the general lack of new plant construction, coal-fired electricity generation and related coal consumption has continued to climb over the past decade. This increase results from utility efforts to optimize performance of existing coal-fired facilities despite their increasing age. Historically, as plants age they become less

[19] Because of regulatory delays, the NO_x requirement was not implemented until 1996.

[20] U.S. EPA, *Acid Rain Program: Technology and Innovation* (May 1996); EPA, 1996 *Compliance Report: Acid Rain Program* (June 1997). Compliance reports for 1997 and 1998 are also available.

reliable and less efficient, leading utilities to derate them and move them from baseload to cycling duties. However, as indicated in **Table** 2, contrary to historical expectations, utilization of coal-fired capacity has increased over the past decade, and the efficiency of units has not decreased.

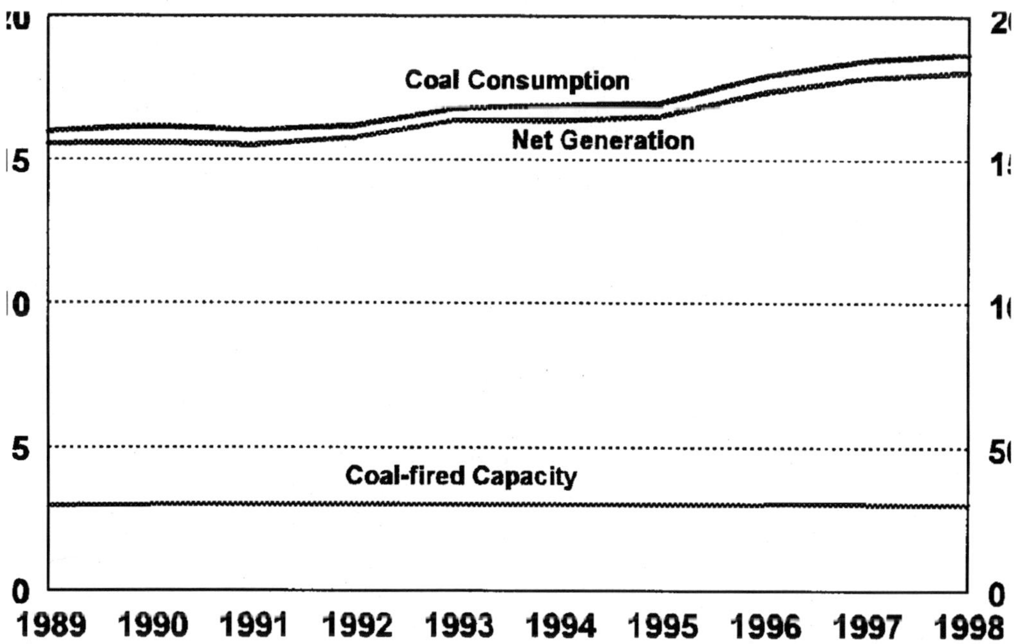

Figure 2. Coal-Fired Capacity, Generation, and Coal Consumption: 1989-1998

Coal consumption in quads

Net generation in trillion kilowatt-hours	Capacity in thousands of megawatt;
20	2000
15	1500
10	1000
5	500
0	0

This suggests that the economics of plant maintenance has changed fundamentally over the past decade or so, making it economic for utilities to spend more to maintain their coal-fired capacity than was the case previously. The question the EPA lawsuits raise is whether these efforts to maintain or even to expand generation from existing coal-fired facilities — compared to the degradation of capacity that would be expected — represent "routine maintenance" or a "major modification" of those facilities under the CAA. If such maintenance does represent a "major modification," then the CAA would require the installation of pollution control equipment; "routine maintenance," on the other hand, would not trigger the requirement for new controls. With the restructuring of the electric utility industry placing ever-greater focus on plant economics, this issue is likely to intensify in the future.[21]

Table 2. Coal-fired Generation Capacity Factors and Heat Rates: 1989-1998 (based on net summer capacity)

Year	Capacity Factor	Heat Rate
1989	59.8%	10,290
1990	59.4%	10,378
1991	59.1%	10,335
1992	59.9%	10,286
1993	62.2%	10,244
1994	62.0%	10,336
1995	62.8%	10,278
1996	65.6%	10,334
1997	67.6%	10,347
1998	68.7%	10,354

Source: Net summer capacity, net generation, coal consumption data from Energy Information Administration, *Annual Energy Review 1998* (July 1999).

WHY NSR?

"There are signs that many utilities will not use the term "life extension" to describe their spending on old power plants, even though extended life is one of the major goals of the spending program.

"The reason for the aversion to the term lies in the 1970 Clean Air Act. That federal law requires all power plants constructed after August, 1971 to restrict emissions of air pollutants such as sulfur dioxide. Plants built prior to 1971 are exempt, which includes most of the early candidates for life extensions. The problem is that the law also says that grandfathered plants can lose their exemption if they are "modified" or "reconstructed " in a major way and emission of proscribed pollutants are increased."[22]

Currently, there is considerable regulatory activity (and judicial response) with respect to electricity generation and air quality. Regulatory activities include the Ozone Transport Rule (also called the NO_x SIP Call), several Section 126 Petitions, a Regional Haze Rule, new NAAQS for Fine Particulates and a revised Ozone NAAQS, a new NSPS for NO_x for electric utilities, initiatives by the Ozone Transport Commission, and several legislative initiatives. Why add NSR enforcement to the list?

As the primary permitting vehicle for stationary sources, NSR is also a primary enforcement tool. Thus, the Department of Justice/EPA lawsuits under NSR represent an attempt to enforce existing laws and regulations, not add new ones. In this sense, the NSR actions are fundamentally different from the activities listed above. Indeed, a different office

[21] See: Larry Parker and John Blodgett, *Electricity Restructuring: The Implications for Air Quality,* CRS Report 98-615 ENR (July 16, 1999).

[22] Robert Smock, Editor, "Power Plant Life Extension Trend Takes New Directions," *Power Engineering (February* 1989), p. 21.

within EPA handles enforcement as opposed to regulatory development. EPA has explicitly stated that the enforcement action under NSR is not linked or tied with the regulatory activities listed above.[23] According to EPA, the action results from an examination of coal-fired utilities that commenced in late 1996.[24]

As an enforcement tool, NSR is attractive because of the remedial actions EPA can seek the court to impose on affected utilities. Besides substantial monetary penalties, EPA can ask the court to require an affected source that violates NSR to install the most recent BACT at its facilities. For SO_x this would involve meeting the percentage reduction requirement promulgated by EPA in 1978. This requirement generally involves installing flue-gas desulfurization (FGD) units (commonly called scrubbers) on coal-fired utility boilers. Likewise, EPA recently promulgated a new NO_x NSPS for new utility boilers; if this were imposed on existing boilers that EPA alleges have been modified, it could require them to install selective catalytic reduction (SCR) units in order to comply. These control devices would reduce emissions by about 70-90% depending on the specific case.

For EPA, these controls could lead to substantial reductions in SO, and NO_x emissions. Noting that the powerplants named in the lawsuits would emit 2.2 million tons of SO, and 0.66 million tons of NO_x in 2000, Administrator Browner stated:

> Controlling the sulfur dioxide and nitrogen oxides from these plants could lead to an 85 to 95 percent reduction respectively in these pollutants. Taken together, these reductions would be the equivalent of taking 26 million cars off the road and reduce acid rain by an estimated 15 percent.[25]

Besides the potential payoff in terms of emissions reductions, the focus on NSR reflects a belief that continuing operation of aging coal-fired plants is not a short-term event, but a wave of the future under electric utility restructuring. As indicated in a previous CRS analysis, extending the life of existing coal-fired capacity is a very cost-effective alternative to constructing new capacity.[26] The more competitive generating market of a restructured industry could further encourage the existing trend to renovate existing coal-fired capacity. One means of mitigating the environmental effects of such a trend would be to require BACT on such facilities through NSR.

---

[23] Steven A. Merman, Letter to Chairman Inhofe (March 26, 1999), p. 3.

[24] Steven A. Merman, Letter to Chairman Inhofe, Answer to Questions #5 and #6 (March 26,1999), p. 14.

[25] Carol M. Browner, Administrator, EPA, "Remarks Prepared for Delivery," Clean Air Enforcement Press Conference, (Washington, D.C.: November 3, 1999), p. 2.

[26] Lany Parker and John Blodgett, *Electricity Restructuring: The Implications for Air Quality,* ORS Report 98-615 (ENR, July 16, 1999).

WHY NOT EARLIER?

"The agency [EPA] is bringing legal actions against the utilities for the very maintenance activities it has approved implicitly for nearly three decades."[27]

The trend toward renovating existing coal-fired capacity is not new, but an increasing trend in the industry for almost two decades.[28] Likewise, the potential emissions impact of such life extension efforts were also recognized early in this time period.[29] However, it wasn't until October 1988 that EPA made a serious attempt to enforce NSR on existing coal-fired electric generating facilities.

This apparent difficulty in enforcing NSR for existing coal-fired facilities could be ascribed to several causes, including EPA budget constraints and priorities. However, the primary substantive difficulty with NSR has been the definition of a major modification. The CAA first used the term "modification" in the 1970 CAA Amendments with respect to the NSPS program. The Act defined modification as "any physical change in, or change in the method of operation of, a stationary source which increases the amount of any air pollutant emitted by such source or which results in the emission of any air pollutant not previously emitted."[30] In subsequent regulations, EPA defined modification as any physical or operational change that resulted in any increase in the maximum hourly emission rate (kilograms per hour) of any controlled air pollutant.[31] In addition, EPA regulations stated that any replacement of existing components that exceeded 50% of the fixed capital costs of building a new facility placed the plant under NSPS, regardless of any change in emissions.[32] With the advent of NSR in 1977, a different approach to defining modification was appropriate as the focus was shifted from enforcing NSPS emission rates to compliance with NAAQS and PSD. In promulgating regulations for the PSD and non-attainment programs, EPA defined "significant" increase in emissions in terms of tons per year emitted by a major source. For sulfur dioxide and nitrogen oxides, the threshold is 40 tons.[33] Facilities that exceed that threshold are subject to NSR.

Enforcing these thresholds has been more difficult than their apparent clarity would suggest. EPA's thresholds for the NSPS program generally represent no practical constraint

---

[27] Edison Electric Institute, *Straight Talk About Electric Utilities and New Source Review* (January 2000), p. II.

[28] As CRS noted in 1985: "However, over the last five years, it has become apparent that the actual lifespan of powerplants is not set, but relatively elastic. With new powerplants costing over $1000 a Kilowatt to construct, utilities have powerful incentives to avoid construction and to rehabilitate older facilities instead. This incentive is partially reinforced by environmental regulations which permit facilities to be rehabilitated up to 50 percent of their assessed value without being required to install NSPS (i.e., scrubbers). With such rehabilitation estimated at about $500 a kilowatt (although that number can vary substantially, operating existing facilities for upwards of 60 years seems to be a developing trend." Larry B. Parker, John E. Blodgett, Alvin Kaufman, and Donald Dulchinos, *The Clean Air Act and Proposed Acid Rain Legislation: Can We Get There from Here?* CRS Report 85-50 ENR (February 21, 1985), p. 46.

[29] See U.S. Congress, Office of Technology Assessment, *Acid Rain and Transported Air Pollutants: Implications for Public Policy,* OTA-0-204 (June 1984), pp. 61-62: Larry B. Parker, John E. Blodgett, Alvin Kaufman, and Donald Dulchinos, *The Clean Air Act and Proposed Acid Rain Legislation: Can We Get There from Here?* CRS Report 85-50 ENR (February 21, 1985), pp. 49-59; Larry B. Parker and Alvin Kaufman, *Clean Coal Technology and Acid Rain Control: Birds of a Feather?* CRS Report 85-207 ENR (October 23,1985), pp. 21-40.

[30] CAA, section 11 l(a)(4), 42 USC 741 l(a)(4).

[31] 40 CFR 60.14(a) (1975).

[32] 40 CFR 60.15(1975).

[33] For PSD, see 40 CFR 52.21(b)(23)(i); for nonattainment, see 40 CFR 52.24(f)(10)

on life extension efforts by utilities. Most life extension efforts improve the availability and reliability of generating units, not their capacity to generate. Thus, their maximum hourly emission rate would not change. Likewise, most life extension efforts cost far less than the 50% asset value threshold.

NSR review has a far more sensitive trigger — a tonnage increase in pollutant output. Because life extension does improve availability and reliability, it is likely to increase emissions over levels emitted before the life extension activities were undertaken. But how does one measure the change? What are the baselines?[34]

These issues came to a head in the late 1980s when EPA decided to enforce NSR against facilities undergoing life extension efforts. In 1988, the EPA ruled that a life extension project by Wisconsin Electric Power Company (WEPCO) met the trigger for NSR because of the potential for increased emissions from the facilities after the project compared with actual emissions from the facilities before the project. After considerable litigation[35] and congressional debate, EPA modified this "actual to potential" emissions trigger with respect to electric utilities in 1992.[36] The new "test" to determine the applicability of NSR compares a facility's actual emissions before the modification with its projected actual emissions after the modification ("actual to future actual"). Specifically, "actual emissions" equal the facility's average emission rate during a 2-year period out of the preceding 5 years before the proposed change. "Future actual" is the product of the facility's projected emission rate after the change and its projected actual utilization based on historical and other data. These are the current NSR regulations for utility plants.

Utilities also responded to EPA's interest in their life extension activities. In defining a modification under NSR, EPA excludes maintenance, repair, and replacement that it considers "routine." In addition, increases in production rates that do not involve capital expenditures do not constitute a modification. Responding to this situation, utilities began to spread out their life extension efforts in an attempt to make them fit into their routine maintenance schedules. Indeed, the term "life extension" has fallen out of the professional literature, replaced with terms like capital improvement, performance improvement and unit integrity, condition assessment, life operation management, review of continued operating requirements, and asset management.[37] The commonly used term currently is rehabilitation program.[38] By spreading out the life extension efforts and integrating them into facilities' operation and maintenance schedules, the distinction between "modification" and "routine maintenance" is effectively blurred, and arguably, eliminated.

---

[34] Defining the baseline has been a key issue. Every powerplant has what is called "nameplate" capacity, which indicates its theoretical size; but the actual output is defined by its "operating capacity," which is determined by the engineering and operational details of the individual plant. Moreover, from an engineering perspective, the operating capacity declines over time as a result of boiler deterioration, pipe clogging, and other predictable changes due to use. The issue is, then, what level of capacity restored by renovations trigger NSR: only renovations that increase capacity beyond the facility's nameplate capacity? those that increase capacity beyond the *original* operating capacity? those that increase capacity above an engineering-defined capacity that projects declines over time? Or those that increase potential emissions above the actual emissions before the modification?

[35] *Wisconsin Electric Power Company v. Reilly,* 893 F.2d 901 (7th Cir. 1990)

[36] 57 *Federal Register* 32314-32339 (July 21, 1992).

[37] Robert G. Presnak and Bock H. Yee, "Life Extension: The Benefits are Real," *Power Engineering* (December 1993), pp. 25-27

[38] For a current view of managing existing facilities, see Jason Makansi, "Rehab: Get the Most from the Existing Asset Base," *Power* (June 1999), pp. 30-40.

WHAT HAS EPA DONE?

"*We expect no early* resolution."[39]

The Justice Department's lawsuits (on behalf of EPA) represent an attempt to enforce the New Source Review requirements of the Clean Air Act with respect to (1) instances of "new construction" of coal-fired capacity by investor-owned utilities and (2) instances in which changing investor-owned utility strategies for managing existing coal-fired facilities become "modifications" that in EPA's view trigger the applicability of the WEPCO rules. In addition, EPA has issued a "Compliance Order" concerning instances of modifications by the Tennessee Valley Authority of its coal-fired power plants that EPA believes should have triggered new source reviews.

Tables 3, 4 and 5 identify the facilities the Justice Department/EPA have cited as violating NSR requirements and briefly note some of the kinds of rehabilitations that EPA views as triggering the "modification" criterion for NSR. These tables are based on the original complaints (see footnotes I and 4). The Justice Department/EPA has said that these complaints may be modified (typically to add new allegations) and that additional complaints may be issued as the agency's investigations proceed.[40]

Table 3. Alleged New Construction Cited in DOJ/EPA Complaints

Facility	Charged with failure to obtain a PSD permit and to apply BACT	Charged with failure to comply with NSPS Subpart Da, e.g., to conduct a performance test
Alabama Power & Southern Company Services, Inc., Miller Unit 3	X	X
Alabama Power & Southern Company Services, Inc., Miller Unit 4	X	X
Georgia Power & Southern Company Services, Inc., Scherer Unit 3	X	X
Georgia Power & Southern Company Services, Inc., Scherer Unit 4	X	X

[39] Statement of David Flanney, Midwest Ozone Group, reported in "Utility Suit to Drag in 2000" *Air Daily* (January 3, 2000), p. I.

[40] E.g., see "Justice Department Amending Complaints in NSR Suit," *Air Daily,* vol. 7, no. 17 (Jan. 25, 2000), 1.

Table 4. Alleged "Modifications" Cited in DOJ/EPA Complaints

Facility	Charged with failure to obtain a PSD permit and to apply BACT	Charged with failure to comply with NSPS Sub-part Da, e.g., to conduct a performance test
Alabama Power and Southern Company Services, Inc., Barry	X (specifically citing "a new design spiral fin economizer in Unit 5")	
Alabama Power and Southern Company Services, Inc., Gorgas	X (specifically citing "(1) a balance draft conversion in 1985; and (2) installation of a new design spiral fin economizer in 1994 and a major upgrade of air heaters in 1994")	
Georgia Power and Southern Company Services, Inc., Bowen	(specifically citing a "new economizer in Unit 2") X	
AEP Service Corp. and Indiana Michigan Power, Tanners Creek Plant	X (specifically citing 5 sets of modifications, including ones involving Units 3, 4, and 5)	
AEP Service Corp. and Ohio Power, Muskingum River Plant	X (specifically citing 8 sets of modifications, which involve Units I through 5)	X (specifically citing 5 modifications, including ones involving Units 3, 4, and 5)
AEP Service Corp. and Ohio Power, Mitchell Plant	X (specifically citing 5 modifications, involving Units I and/or 2)	
AEP Service Corp. and Ohio Power, Cardinal Plant	X (specifically citing 7 modifications, involving Units I and/or 2)	X (specifically citing 5 modifications, involving Units I and/or 2)
AEP Service Corp., Appalachian Power, Ohio Power, and/or Central, Philip Spom Plant	X (specifically citing 5 modifications, including ones involving Units I through 5)	
PSI and/or Cinergy, Cayuga Plant	X (specifically citing 5 modifications, including ones involving Units I and/or 2)	X (specifically citing the replacement of economizers on Units I and 2)
CG&E and/or Cinergy, Beckjord Plant	X (specifically citing 5 modifications, including ones involving Units I through 5)	
Southern Indiana Gas and Electric Company, Culley Station	X (specifically citing 4 modifications, involving Units 1, 2, and 3)	X (specifically citing "replacement activities" at Unit 3)
Illinois Power Company, Baldwin Power Station	X (specifically citing 7 modifications, involving Units 1, 2, and/or 3)	X (specifically citing "replacement activities" at Units I and 2)
Ohio Edison Company, Pennsylvania Power Company, subsidiaries of FIRSTENERGY Corp., Sammis Station	X (specifically citing 11 modifications, one or more involving each of Units I through 7)	X (specifically citing replacement of Unit 6's burners in 1992 and of its coal pulverizers in 1998)
Tampa Electric Company, Big Bend	X (specifically citing replacement of steam drum internals in Unit I in 1994 and in Unit 2 in 1991; and the waterwall, and high temperature re-heater in Unit 2 in 1994)	
Tampa Electric Company, Gannon	X (specifically citing 1996 furnace floor replacement in Unit 3 and 1994 cyclone replacement in Unit 4)	

Table 5. Tennessee Valley Authority: Compliance Order

TVA Facility	Charged with failure to obtain a PSD permit	Charged with failure to comply with NSPS Sub-part Da, e.g., to conduct a performance test
Paradise. Units 1. 2. & 3	X (specifically citing replacement of the cyclones, lower furnace walls, and floor in 1985)	X, Paradise Unit 3
Colbert Unit 5	X (specifically citing re-habilitation and modification of boiler, turbine, and controls in 1982)	X
Widows Creek Unit 5	X (specifically citing the replacement of the reheater and secondary superheater crossovers and elements in 1989)	
Allen Unit 3	X (specifically citing the replacement of the reheater in 1991-2)	
John Sevier Unit 3	X (specifically citing the replacement of all waterwall and burner wall tubes, and of superheater platen elements in 1988)	
Cumberland Units 1 & 2	X (specifically citing the replacement of front and rear secondary super-heater outlet headers and of the inlet terminal tubes and main steam piping tee in 1993)	
Bull Run Unit I	X (specifically citing the replacement of the secondary superheater outlet pendant elements and of all economizer elements in the "A" and "B" furnace in 1987)	

Basically, each of the complaints against investor-owned facilities seeks injunctive relief and civil penalties of up to $25,000 per day of violation. The injunctive relief typically calls for the defendants to comply in the future with all CAA requirements; and to remedy any past NSR violations by installing BACT as appropriate, to apply for permits, and to audit operations to assure compliance with all NSR requirements. For TVA, EPA's compliance order asks TVA to develop a schedule for meeting all applicable requirements affecting the facilities specified, to audit all its coal-fired power plants to determine NSR compliance and to develop a schedule for remedying any noncompliance discovered during the audit, and to retire SC) 2 allowances equal to any SO, reductions that result from complying with the order.

The industry's response to EPA's enforcement action has focused on an exemption under NSR for "routine maintenance." NSR regulations state that a major modification does not include "routine maintenance, repair, and replacement.'"[41] The industry argues that although EPA has never explicitly defined this phase, EPA has over the past 20 years not objected to utilities engaging in those very activities that are now being cited in the complaints as triggering NSR. Industry further asserts that these activities were well-publicized in journal articles and government reports.[42] In effect, industry is charging EPA with changing the rules in the middle of the game. On the substance of the issue, the industry argues that the activities cited as violating NSR are nothing more than those necessary to meet its responsibilities to provide for the reliable, safe, and efficient operation of its plants through proper maintenance practices — analogous to replacing worn tires on a car. That these activities might serve to extend the life of the powerplants is coincidental, and thus should not trigger the major modification criterion of NSR.

WHAT COULD THE NSR ENFORCEMENT ACTION ACCOMPLISH IN TERMS OF EMISSION REDUCTIONS?

"Emissions cannot exceed the cap.
"Emissions won't be below the cap either - allowances not used at one source will be emitted somewhere else.
"If you still apply NSR to sources under the cap, other sources will produce the pollution avoided by the source subject to NSR. This will all occur at a higher cost. "[43]

In announcing the NSR suits, the EPA Administrator states that "controlling the sulfur dioxide and nitrogen oxides from these plants could lead to an 85 to 95 percent reduction respectively in these pollutants."[44] Based on her statement, this would reduce 802 emissions by 1.87 million tons and NO_x emissions by 0.63 million tons. (Compared to the slippage of title IV reductions noted earlier, these cuts in emissions would more than compensate for the erosion in the original title IV reductions: for 802, by about 50% more than the slippage, and for NO_x, by about double the slippage.) Also, given the widespread nature of life extension efforts, it is reasonable to assume that further reductions would be achieved as other utilities either installed BACT or retired their offending facilities. Thus, at first glance, it would appear that very substantial emission reductions could be achieved by EPA's NSR action.

However, the CAA is a complex piece of legislation built up over time. In the case of 802 any reduction achieved under NSR would interact with reduction requirements under title IV — a 802 reduction program designed with different premises than NSR. Specifically, title IV limits total S02 emissions from utilities to 8.9 million tons beginning in the year 2000, with interim reductions required in 1995. The cap is enforced through tonnage limitations at individual existing utility plants and by an emission offset requirement for new facilities. SO_2 emissions from most existing sources are capped at a specified emission rate times a historical (1985-1987) average fuel consumption level. Thus the tonnage limitation is based on preset

[41] 40 *CFR* 52.21(b)(2)(iii)(a)

[42] For a review of industry's position, see Edison Electric Institute, *Straight Talk About Electric Utilities and New Source Review,* Edison Electric Institute (January, 2000).

[43] Environmental Protection Agency, OAQPS, "An Emissions Cap Alternative to New Source Review" (September 27, 1999), p. 8.

[44] Carol M. Browner, Administrator, Remarks Prepared for Delivery, Clean Air Enforcement Press Conference (Washington, D.C.: November 3, 1999).

and historical data, not regulatory limits. To implement the program, title IV created a comprehensive emissions allowance system. An allowance is a limited authorization to emit a ton of SO_2 during or after a specified year. Issued by EPA, allowances are allocated to existing facilities in accordance with the emission rate/fuel consumption formulas detailed in the law. Such allowances may be used at the plant they are allocated to, or be traded or banked for future use or sale. The program has been very successful with 100% compliance.

This 1990 CAA Amendments program does not integrate well with the 1977 CAA Amendments NSR program. Except that they both focus on existing facilities and S02, they have little in common. The NSR is concerned with modifications at existing facilities and installation of BACT. Title IV doesn't address whether existing facilities continue operation or not, or whether a specific facility installs BACT or not; compliance with the cap is the determining criterion. NSR is an enforcement mechanism to assure compliance with individual plant standards; title IV is a program to reduce aggregate SO_2 emissions by permitting utilities considerable flexibility in determining appropriate compliance strategies.

The current SO_2 NSPS, the "floor" for any BACT or LAER determination, is a percentage reduction requirement that reduces SO_2 emissions by 70%-90%, depending on the coal burned. However, the allocations under title IV for existing coal-fired facilities is not as stringent and can be met with low-sulfur coal. Thus, any facility that installed BACT under NSR would "overcontrol" SO_2 under title IV, and, therefore, have excess allowances available for sale or to bank for future sale. Because of this, any reductions achieved because of NSR enforcement could be rendered moot by title IV, if the affected plant subsequently sold its SO_2 reduction to some other facility not covered by a NSR action.[45] Except for any TVA reductions, the net result would be no reductions, at least theoretically. There is also no provision in title IV for adjusting allowance allocations as a result of NSR enforcement. Indeed, the law is explicit in that its allowance allocations are based on historical data, not on any presumption of compliance with NSPS or SIP requirements. To avoid this "allowance trap," either Congress would have to change the law, or utilities would have to voluntarily agree to surrender the excess allowances created by the NSR action.

The situation with potential NO_x reduction is more complex. First, there is the interaction of NSR and the NO_x NSPS. Unlike the very stringent SO, NSPS, the NO_x NSPS historically has not reflected the cutting edge in technology development.[46] Until the new standard was set in 1998, the NO_x NSPS for coal-fired facilities was 0.6/0.5 lb. of NO_x per million Btu of heat input, depending on the type of coal burned. This standard, set in 1979, could be met with fairly simple combustion modifications or low- NO_x burners, and did not require the installation of pollution control devices such as selective catalytic reduction (SCR). Indeed, the standard did not reflect the state of the art with respect to low-NO_x burners.

In 1998, EPA promulgated a new NO_x NSPS for coal-fired facilities of 0.15 Ib. of NO_x per million Btu — a standard more in line with available technology. However, this new standard was challenged in court. In September, 1999, the D.C. Court of Appeals vacated the new NO_x NSPS with respect to modified utility boilers, while upholding the NSPS with

[45] The TVA Compliance Order would require retirement of allowances equal to any SO, reductions achieved as a result of the compliance order.

[46] Larry Parker, *Nitrogen Oxides and Electric Utilities: Revising the NSPS,* CRS Report 96-737 ENR (October 13, 1998).

respect to new sources.[47] By vacating the modified standard, the NSPS for modified sources returns to the previous 1979 standard until such time as EPA proposes a revised NSPS. As a result, the floor for determining BACT or LAER for modified coal-fired sources is unclear at the current time. If the floor is the current modified NSPS as set in 1979, reductions achieved by NSR enforcement would be considerably less than anticipated by EPA. In contrast, if the floor is the new 1998 NSPS, the reduction would be substantial, as suggested by EPA.

The confusion is exacerbated by the interaction of NSR and title IV. The NO_x reduction program under title IV differs substantially from the SO_2 program. Like the NSPS program, the title IV NO_x program is based on emission rates, not tonnage limitations. The difference is that the emission rate for the title IV program is set for existing facilities to be achieved in 1995 or 2000 (depending on the facility), regardless of whether the facility is modified or not. In addition, the rate limitation for most boilers under title IV is 0.45 to 0.5 lb. per million Btu, or more stringent than the 1979 NO_x NSPS. Thus, you have the curious situation of some existing coal-fired facilities having emission controls since 1995 that are more stringent than the existing NSPS — a situation that continues currently with the court action on the modified NSPS.

Surveying BACT determinations over the time period 1991-1995 sheds no light on what BACT might be currently: data indicate permitted emission rates ranged from 0.15 to 0.5 lb. per million Btu.[48] Thus, it is difficult at the current time to project what any actual NO_x reduction would be achieved by EPA's NSR action.

WHAT ARE THE ALTERNATIVES?

"EPA Offers Utilities Off-Ramp From NSR 'Highway to Hell'"[49]

By seeking to enforce NSR requirements, EPA exploits an existing authority that holds the potential for reducing emissions. EPA clearly believes that some sources have been evading statutory requirements. However, according to some, the action appears to raise the classic enforcement issue: will the outcome be reduced emissions, or just costly litigation consuming agency and utility resources that might be more effectively invested in other pollution controlling activities?

NSR was one approach that the Clean Air Act took to control emissions from existing sources, but arguably more efficient and more effective methods to ensure declining emissions from existing sources over time have been developed since NSR provisions were added to the CAA in 1977. For example, title IV of the CAA, enacted in 1990, explicitly and substantially reduces SO_2 and NO_x emissions from existing utility plants. In fact, title IV reduced more SO_2 emissions from coal-fired electric generating facilities in its first year of implementation (1995) than NSR has in its 20 years of existence. The "cap and trade" program has had 100% compliance (indeed, substantial over-compliance); the implicit logic of EPA's lawsuits suggests NSR's compliance has been near zero. The title IV program began

[47] *Lignite Energy Council v. Environmental Protection Agency,* Order No. 98-1525, D.C. Court of Appeals (September 21, 1999).

[48] Office of Air and Radiation, EPA, *Analyzing Electric Power Generation under the CAAA* (July 1996).

[49] Chris Holly, "EPA Offers Utilities Off-Ramp from NSR 'Highway to Hell,'" *The Energy Daily* (January 14, 2000), p. I [headline].

without significant delays (SO, program on-time, NO_x program I year late); the EPA lawsuits could take years to resolve with uncertain results.

A first alternative is to expand and build on the success of title IV's cap and trade program. If the object of NSR enforcement is to reduce SO, emissions from coal-fired powerplants, the most straightforward alternative would be to lower the cap on such emissions contained in title IV. The practical effect of the 1990 SO, cap was to reduce SO, emissions from existing facilities to the level required by the 1971 NSPS. The effect on new sources was to reduce the NSPS to zero, as all emissions now have to be offset. Lowering the existing cap by about two-thirds would achieve roughly the same emission reductions as all existing powerplants meeting the 1978 NSPS, but utilities would have some flexibility in achieving such reductions. Admittedly, utilities would get credit for shutdowns that they would not get under NSR enforcement; however, the administrative and cost advantages of the allowance system might be considered worth it. In any case, it would require new legislation, which could be a long drawn out process.

Similarly, a new "cap and trade" program for NO_x would eliminate the uncertainties involved in the NSR enforcement debate, and, potentially, in several other EPA initiatives with respect to NO_x emissions. Not surprisingly, EPA has been strongly urging states to consider a regional cap and trade program in implementing its currently stalled Ozone Transport Rule, and any possible compliance with Section 126 petitions. Indeed, EPA has gone so far as to make such a program a part of its proposed Federal Implementation Plan (FIP) for states that do not submit adequate SIPs under the OTR and its compliance plan for implementing approved Section 126 petitions. However, to implement a regionwide cap and trade program under the stalled NO_x SIP Call for NO_x would require either extraordinary cooperation between the states affected (because of the SIP process), or new EPA authority.

A second alternative to NSR is to focus on SO, and NO_x emissions through other regulatory initiatives. There are several regulatory initiatives underway at EPA that would reduce SO, and/or NO_x emissions in the future. Besides the previously mentioned OTR and Section 126 petitions, EPA has promulgated a fine particulate NAAQS and a regional haze rule. Both of these regulations will require reductions in SO, and NO_x emissions as they are implemented. The problem with them is that their implementation schedules are very long, with few reductions occurring before the year 2010. Unlike the alternative of tightening the title IV SO, cap and creating an NO_x cap, these initiatives do not require further legislative authority.

A third alternative to NSR would be to initiate controls on other pollutants associated with coal-fired powerplants, such as mercury (Hg) and carbon dioxide (CO,) — controls that would correlatively tend to reduce other emissions, such as SO_2 and NO_x, as well. Currently, CO, is not considered an air pollutant under the CAA. Mercury is a hazardous air pollutant under the CAA, but regulation of utility boilers is contingent upon an EPA determination of its appropriateness and necessity.[50] As a result, the process to begin controlling Hg emissions from electric utility boilers has moved very slowly. EPA has argued that it has authority under the CAA to regulate CO, however, that opinion is very controversial.[51] In lieu of a potentially lengthy legal battle, any CO, reductions in the near

[50] Clean Air Act, section 112(m)(l).

[51] Jonathan Z. Cannon, General Counsel, *EPA's Authority to Regulate Pollutants Emitted by Electric Power Generation Sources,* Memorandum to Carol M. Browner (April 10,1998).

term would have to result from either voluntary initiatives or through ratification of the Kyoto Protocol.

A final alternative is to reform NSR. One reform could be new legislation to define "modification" in terms of plant age, not physical or operation change. For example, the definition of modification could be altered to include any electric utility generating facility that was 40 years or older, regardless of whether it had been "modified" or not. It would simplify the NSR trigger, but would required legislative action.

Another reform could emerge from a current EPA initiative. It is not surprising, given the situation outlined above, that EPA has been considering reforms to NSR for several years. One focus of that effort has been proposed "off-ramps" to NSR. One off-ramp, called the "Clean Unit Exclusion," would exclude existing units from the complicated NSR applicability rules where they already have state-of-the-art pollution control technology.[52] This exclusion would recognize that NSR is unlikely to result in significant emission reductions if units are already well-controlled. In the proposal, a "clean unit" is defined as one that has installed BACT or LAER within the last 10 years. Such units could be modified in any fashion that would not increase the unit's maximum hourly emissions unit. However, how much of an improvement this is over the current situation is debatable.

From EPA's perspective, the problem with all the alternatives is their contingency: they imply the need for legislation or for finalization of regulations, which require time and which may be problematic. Of course, while EPA has the existing authority to initiate its NSR suits, the outcome of the litigation is also problematic.

CONCLUSION: NSR — AMBIGUOUS, MEANINGLESS OR MOOT?

"EPA is embracing cap and trade programs for many pollutants.
"NSR is duplicative and counterproductive on top of a cap.
"Replacing NSR with a cap provides the same environmental protection with lower cost and greater environmental benefit."[53]

Much of the popular debate on NSR has focused on "grandfathered" powerplants. One example from a 1998 report by a public interest group states:

The Clean Air Act of 1970, as amended in 1977 and 1990, contains a major exemption that allows older coal-burning power plants to emit between 4 and 10 times the amount of pollution that new plants may emit under the Clean Air Act. In part, this colossal loophole exists because industry lobbyists argued successfully that its older plants would soon retire, and that therefore it would be wasteful to require expensive retrofits to control pollution from these plants. However, over 20 years later, many of these same plants, built in the 30s, 40s, 50s and 60s, are still operating, largely without environmental controls.[54]

The term "grandfathered powerplant" is a much used and little understood concept employed in debate on emissions from existing powerplants. Specifically, "grandfathered" is an ambiguous, and, in some cases, meaningless term generally used to indicate whether a given powerplant is covered under Section I II of the Clean Air Act. Passed with the 1970

[52] 61 *Federal Register* 38255 (July 23, 1996).

[53] Environmental Protection Agency, "An Emissions Cap Alternative to New Source Review" (September 27, 1999), p. 2.

Clean Air Act Amendments, Section 111 requires the EPA to promulgate regulations defining the minimum controls necessary for new sources (including power plants) regardless of their location. Called New Source Performance Standards (NSPS), they require major new sources constructed after their promulgation to install the best system of continuous emission reduction which has been adequately demonstrated according to EPA. Currently, there are NSPS regulations for powerplants that cover three pollutants — sulfur dioxide (SO_2), nitrogen oxides (NO_x), and particulate matter. Typically, "grandfathered" refers to those plants (usually coal-fired powerplants) that were constructed before the effective dates of those NSPS regulations and, hence, not subject to them. NSPS regulations for powerplants were first promulgated in 1971 and revised in 1979. The NO_x NSPS regulations for powerplants were revised again in 1998. Instead of NSPS requirements, such "grandfathered" sources must meet emission rate limits established by a State Implementation Plan (SIP).

Three aspects of the NSPS make the term "grandfathered" at best ambiguous, and at worst, meaningless.

Some pollutants of concern, such as carbon dioxide (CO,) and mercury (Hg) are not criteria air pollutants, and, therefore, not covered by the NSPS for powerplants at the current time. Hence, "grandfathered" is meaningless for these pollutants as all powerplants (indeed, all major sources of these pollutants) are "grandfathered."

EPA is required to review the NSPS every eight years, resulting in increased stringency for covered pollutants as technology improves (and for determinations of BACT and LAER). Therefore, what powerplants are "grandfathered" is ambiguous as there is no set baseline. For example, the NSPS for NO_x was revised in 1998 to a stringency that only a couple of commercially operating coal-fired powerplants currently met; by that standard, virtually all coal-fired powerplants are "grandfathered" with respect to NO_x emissions.

Title IV of the 1990 Clean Air Act Amendments changed the regulation of existing powerplants with respect to SO, and NO_x. In some cases, the requirements under title IV for existing sources are more stringent than some of the existing or previous NSPS. For example, under phase I, some existing "grandfathered" powerplants were required in 1995 to meet NO_x standards more stringent than then existing NSPS NO_x requirements for new powerplants. Likewise, under phase 2 of title IV, existing "grandfathered" coal-fired powerplants are required in 2000 to meet SO, standards that are essentially equivalent to (if not more stringent than) the 1971 NSPS for SO_2. The term "grandfathered" is essentially meaningless under such circumstances. If the focus of debate about "grandfathered" powerplants is NO_x emissions, then age of plant is not a relevant consideration — fuel source is. Coal-fired facilities, regardless of age, are the relevant focus of any effort to increase NO_x controls. If the focus of debate about "grandfathered" powerplants is SO, emissions, then the title IV emissions cap is the relevant consideration. There, age was a consideration in allocating emission credits; however, the relevant definition was not based on NSPS compliance (or any other CAA compliance), but on whether the plant was operational, under construction, or planned at the time of enactment. Indeed, the NSPS for SO_2 for new powerplants is in some ways moot - all new sources must completely offset their emissions under the cap as they receive no allocation of emission credits. The NSPS is effectively nil. Thus, if reducing SO, from electric generating facilities is the goal, shrinking the current cap on SO, is the most logical approach. Likewise, a cap on NO_x emission is a logical extension for reducing NO_x emissions from

[54] United States Public Interest Research Group, *Lethal Loophole,* U.S. PIRG Education Fund (June, 1998), p. 3.

electric generating facilities. Not coincidentally, EPA favors this approach in addressing transported pollution programs in the Northeast, and the agency has proposed state-by-state emissions caps.

The WEPCO decision precipitated public debate and congressional oversight, and the DOJ/EPA actions on NSR may do likewise. Many of the same issues are being raised: The suits depend on interpretations of the legislative intent of several CAA provisions, especially with respect to modifications of existing sources. Also, many alternatives for addressing NSR issues would require new legislation amending the CAA, in order to bring consistency to the NSR provisions and to update the Clean Air Act's approach to regulation of the electric utility industry's environmental effect to reflect the knowledge gained by 30 years of CAA implementation.

Chapter 7

ELECTRICITY RESTRUCTURING AND AIR QUALITY: COMPARISON OF PROPOSED LEGISLATION*

Larry B. Parker

INTRODUCTION

Electric utility generating facilities are a major source of air pollution. The combustion of fossil fuels (petroleum, natural gas, and coal), which account for 67% of U.S. electricity generation, results in the emission of a stream of gases. These gases include several pollutants that directly pose risks to human health and welfare, including sulfur dioxides (SO_2), nitrogen oxides (NO_x), and mercury (Hg). Other gases may pose indirect risks, notably carbon dioxide (CO_2), which may contribute to global warming.[1] Table 1 provides estimates of SO_2, NO_x, and CO_2 emissions from electric generating facilities. Emissions of Hg from utility facilities are more uncertain, and are estimated at about 52 tons. Utilities are currently subject to an array of environmental regulations, which differentially affect both the cost of operating existing generating facilities and of constructing new ones.

Table 1: Emissions From U.S. Fossil-fuel Electric Generating Plants (thousands of short tons)

Emissions	1994	1995	1996	1997	1998
SO_2	14,211	11,437	12,053	12,317	12,432
NO_x	6,790	6,737	6,996	7,227	7,221
CO_2	1,986,079	1,995,471	2,065,339	2,142,118	2,209,286

Source: Energy Information Administration, *Electric Power Annual 1998*, vol. II, p. 40

* Excerpted from CRS Report RS20326. Updated July 26, 2000.

[1] Steam-electric utilities produce only minor amounts of volatile organic compounds (VOCs), carbon monoxide (CO), and lead — on the order of 2% or less of all sources.

After many decades of operating in a comprehensive regulated market structure, the electric utility industry is facing significant change, both from new generating and transmission technology and from shifting policy perspectives with respect to competition and regulation. At issue is whether these changes will increase air pollution emissions. Specifically, some argue that this singular focus on economic efficiency could come at the expense of other values that the regulatory system traditionally has balanced against economic efficiency, particularly equity and environmental considerations. The environmental concern with respect to restructuring is that the new economic signals being given by a competitive generation market could result in increased emissions of undesirable pollutants, particularly NO_x.

Previous CRS analysis suggests that the environmental effects of restructuring depend on whether, for conventional air pollutants, the existing regulatory regimen will work effectively as the industry structure changes. For some pollutants, such as SO_2, the nationwide emission "cap" seems secure; but for others, particularly NO_x, the state-led implementation process may have difficulty coping with regional disparities in emissions. For CO_2, any controls would be contingent on future ratification of the Kyoto Agreement to curtail emissions and on domestic implementing legislation.[2]

PROPOSED LEGISLATION IN THE 106TH CONGRESS

Currently, ten bills have been introduced in the 106th Congress to increase pollution controls on electric generating facilities. They are summarized in table 2. All of the bills control at least NO_x and SO_2; others include CO_2 and Hg. *S. 673*, was not included as it pertains only to Hg, and its scope extends beyond electric generating facilities.

Two of the bills, *H.R. 25* and *S. 172* are companion legislation focused on SO_2 and NO_x, with EPA to regulate Hg after a monitoring network has been installed. The bills build on the SO_2 allowance trading scheme contained in title IV of the 1990 Clean Air Act Amendments (CAAA); under this program utilities are given a specific allocation of permitted emissions (called allowances), and may choose to use those allowances at their own facilities, bank them for future use, or sell them to other utilities needing additional allowances. A third bill, *H.R. 657* is very similar to *H.R. 25*, *S. 172* except that the penalties for non-compliance with the NO_x provisions are more severe.

S. 1369 and *H.R. 2569* also have similarities in terms of controlling SO_2, NO_x, CO_2 and Hg, but have significant differences, particularly with respect to implementation strategies. *H.R. 2569*, like *H.R. 25*, *S.173* and *H.R. 4861*, builds on the allowance system contained in title IV of the 1990 CAAA. In contrast, *S. 1369* uses a credit system, as used in EPA's offset program, to allow facilities to gain credits from reducing their emissions below that specified in the legislation. Thus, a utility would get an emission credit after it has shown its over-compliance with the bill's provisions. How this *post facto* system would work with the allowance type system of title IV is unclear.[3]

---

[2] Larry Parker and John Blodgett, *Electricity Restructuring: The Implications for Air Quality*, CRS Report 98-615, updated July 14, 2000.

[3] For more on the difference between allowance type and credit type schemes, see: Larry Parker, *Global Climate Change: Market-Based Strategies to Reduce Greenhouse Gases*, CRS Issue Brief IB97057, updated regularly.

The sixth bill introduced, *H.R. 2645*, is the most stringent with respect to mandating "elimination" of Hg emissions. The bill is also unique in employing an emission rate approach to implementation. Utilities would be required to meet a specified emission rate, rather than a tonnage cap as the above bills employed. The rate would be based on the desired emission cap specified in the bill and an EPA estimate of expected electricity generation in a given year.

The seventh bill introduced, *H.R. 2900*, controls SO_2, NO_x, CO_2, and Hg, but provides considerable flexibility to EPA in developing implementation strategies. Specifically, EPA implementing regulations are required to allocate reductions equitably and may consider market-oriented mechanisms (except for Hg).

The eighth bill introduced, *H.R. 2980*, controls SO_2, NO_x, CO_2, and Hg, but with individual unit-by-unit requirements for SO_2 and NO_x based on output-based emission rates and average 1996-1998 fuel consumption, a percentage reduction requirement for Hg, and an allowance based system for CO_2. Apparently, plants built after 1998 must emit no SO_2 or NOx, as their historical fuel consumption would be zero.

The ninth bill introduced, *S. 1949*, is unique in combining individual unit-by-unit emission rate limitations and/or percentage reduction requirements with mandated combustion efficiency standards to control SO_2, NO_x, CO_2, and Hg. The bill does not include any sort of allowance or credit trading program.

The tenth bill introduced, *H.R. 4861*, is an expansion and strengthening of *H.R. 25*. Besides extending controls to include CO2, the bill includes a renewable portfolio standard and net metering provisions.

Table 2: Comparison of NOx Control Proposals

Bills Provisions	***H.R. 25*** (Boehlert)/ ***S. 172*** (Moynihan) and ***H.R. 657*** (Sweeney)	***S. 1369*** (Jeffords)	***H.R. 2569*** (Pallone)	***H.R. 2645*** (Kucinich)	***H.R. 2900*** (Waxman)	***H.R. 2980*** (Allen)	***S. 1949*** (Leahy)	***H.R. 4861*** (Lazio-Boehlert)
Emissions Cap on NO_x	estimated at 2.36 million tons in 2005 with interim reductions	1.66 million tons in 2005	1.66 million tons in 2005 with interim regional reductions	1.66 million tons in 2005	estimated at 1.8 million tons in 2005	estimated at 1.6 million tons in 2005, declining with plant retirements	estimated by sponsor at 1.4 million tons within 10 years of enactment	estimated at 2.13 million tons in 2005
Emissions Cap on SO_2	4.45 million tons in 2005	3.58 million tons in 2005	4.0 million tons in 2004	3.58 million tons in 2005	estimated at 3.11 million tons in 2005	estimated at 3.2 million tons in 2005, declining with plant retirements	estimated by sponsor at 2.9 million tons within 10 years of enactment	2.225 million tons in 2005
Emission Cap on CO_2	not covered	1.914 billion tons in 2005	1.914 billion tons in 2005	1.71 billion tons in 2005	estimated at 1.914 billion tons in 2005	1.914 billion tons in 2005	estimated by sponsor at 1-1.35 billion tons within 10 years of enactment	1.914 billion tons in 2005
Emissions Cap on Mercury	EPA to regulate	5 tons in 2005	estimated at about 5 tons in 2010 with interim reductions	"elimination" by 2010	estimated at about 5 tons in 2005	estimated at 15 tons in 2005	estimated at 5 tons within 10 years of enactment	5 tons in 2005

Bills Provisions	*H.R. 25* (Boehlert)/ *S. 172* (Moynihan) and *H.R. 657* (Sweeney)	*S. 1369* (Jeffords)	*H.R. 2569* (Pallone)	*H.R. 2645* (Kucinich)	*H.R. 2900* (Waxman)	*H.R. 2980* (Allen)	*S. 1949* (Leahy)	*H.R. 4861* (Lazio-Boehlert)
Scope	48 contiguous states and DC	50 states and DC	48 contiguous states and DC; interim reductions apply to 22 eastern states	50 states and DC	50 states and DC	50 states and DC	50 states and DC	48 contiguous states and DC
Affected Units	electric generating facilities 25 Mw or greater	electric generating facilities 15 Mw or greater	electric generating facilities 15 Mw or greater	electric generating facilities 15 Mw or greater	electric generating facilities 15 Mw or greater	electric generating facilities 15 Mw or greater (50 Mw for CO_2)	all "electric utility generating units"	electric generating facilities 25 Mw or greater
Penalties for non-compliance	NOx: $6,000 per excess ton plus one-for-one offset from future emission allocations. For *H.R. 657*, $12,000 per excess ton plus two-for-one offset SO_2: same as CAA, title IV	NOx: $15,000 per excess ton SO_2:$2,500 per excess ton CO_2: $100 per excess ton Mercury:	all pollutants include a one-for-one offset, plus: NOx: $5,000 per excess ton SO_2: EPA determines CO_2: $100 per excess ton	$100,000 each day a facility exceeds the specified emissions rate as determined by EPA	determined by EPA	NOx, SO_2, Hg: no special penalties specified - CAA penalties would apply CO_2: $100 per ton plus one-for-one offset from future emission allocations	no special penalties specified - title V permits required	NOx: $6,000 per ton plus one-for-one offset from future emission allocations SO_2: same as CAA, title IV CO_2: $100 per ton plus one-for-one offset from future emission allocations
Special Provisions	NOx allowance value halved during ozone	NOx plant allocation	NOx: reductions	EPA to establish	all powerplants	permanent CO_2 and NOx	all powerplants	NOx allowance

Bills Provisions	*H.R. 25* (Boehlert)/ *S. 172* (Moynihan) and *H.R. 657* (Sweeney)	*S. 1369* (Jeffords)	*H.R. 2569* (Pallone)	*H.R. 2645* (Kucinich)	*H.R. 2900* (Waxman)	*H.R. 2980* (Allen)	*S. 1949* (Leahy)	*H.R. 4861* (Lazio-Boehlert)
	season; reserve of allowances for new sources	weighed for the ozone season	increased for facilities emitting above a specific level during ozone episodes citizen suit enforcement provision	schedule of reductions, beginning in 2002	30-years or older must meet current New Source Performance Standard (NSPS) requirements	reductions through plant retirements should be credited in any future climate change implementation program enacted by Congress	subject to NSPS New Source Review requirements within 10 years of enactment all powerplants subject to combustion efficiency standards 10 years after enactment	value halved during ozone season reserve of allowance for new sources renewable portfolio standard of 6% in 2010 with interim targets net metering provision
Implementation Strategy	tradeable allowance system	tradeable emission credits created *post-facto*	tradeable allowance system	emissions rate approach based on EPA's estimate of annual electricity generation	to be determined by EPA — market mechanisms permitted (except for Hg)	unit-by-unit compliance with SO_2, NOx, Hg provisions; tradeable allowance system for CO_2	unit-by-unit compliance	tradeable allowance system

Unless otherwise noted, estimates by CRS using Department of Energy and Environmental Protection Agency data.

Chapter 8

AIR QUALITY: EPA'S OZONE TRANSPORT RULE, OTAG, AND SECTION 126 PETITIONS – A HAZY SITUATION?*

Larry B. Parker and John E. Blodgett

BACKGROUND

Interstate transport of air pollutants is not a new problem. However, the 1990 Clean Air Act Amendments (CAAA) provided the Environmental Protection Agency (EPA) and the states with new tools to address the problem. One of those tools is section 176A, a provision that permits the EPA, either on its own or by petition from any state, to establish a transport region to address regional pollution problems contributing to violations of a primary National Ambient Air Quality Standard (NAAQS). A commission of EPA and state officials is constituted to make recommendations to EPA on appropriate mitigation strategies. Based on the commission's findings and recommendations, EPA is required under section 110(k)(5) to notify affected states of inadequacies in their current state implementation plans and to establish deadlines (not to exceed 18 months) for submitting necessary revisions (commonly known as a SIP call).

Besides authorizing administratively-created transport regions, the 1990 CAA Amendments statutorily created an Ozone Transport Region (OTR) in the Northeast (section 184). This provision required specific additional controls for all areas (not only nonattainment areas) in that region, and established the Ozone Transport Commission (OTC) for the purpose of recommending to EPA regionwide controls affecting all areas in the region.

In addition, the 1990 CAA Amendments rewrote the provisions of Sec. 126 regarding interstate transport of pollution. Under Section 126(b), any state or political subdivision may petition EPA for a finding that any major source or group of stationary sources located in another state are emitting pollutants that "significantly contribute" to the nonattainment of a NAAQS by their state. EPA is to respond to the petition within 60 days. If the petition is granted, the offending sources must cease operations within 3 months unless the sources comply with emission controls and compliance schedules as determined by EPA to bring them into compliance with the section.

THE OZONE TRANSPORT ASSESSMENT GROUP

In 1995, EPA issued a planning guidance memorandum for ozone nonattainment areas classified serious or above. Among its provisions was a call for a 2-year regional assessment of ozone transport and control strategies. The goal was to reach consensus between EPA and the affected states on necessary additional regional and other emission reductions to achieve attainment in ozone transport areas. Based on results, states would submit plans by mid-1997 for appropriate local and regional controls to achieve attainment. If no consensus were reached, EPA stated it would use its authority under the Clean Air Act (CAA) to ensure the required reductions are achieved (particularly Section 126 and/or section 110).[1]

Following a recommendation of the Environmental Council of the States (ECOS) for a national work group on ozone, EPA, the 37 easternmost states and the District of Columbia, industry representatives, and environmental groups formed an Ozone Transport Assessment Group (OTAG) partnership, which initially met in May, 1995. The policy arm of OTAG consisted of the state environmental commissioners from the 37 states and the District of Columbia and senior EPA officials. It was this Policy Group that voted on the proposals to be included in recommendations to the states and EPA.

To develop the necessary assessment data, OTAG created three subgroups and six workgroups to address modeling and assessment issues, strategies and controls, along with outreach and communications. Each subgroup held open working meetings to receive input from various stakeholders. The recommendations of these subgroups were submitted to the Policy Group via joint sessions of the Strategies and Controls and Modeling and Assessment Subgroups. The final OTAG meeting was June 19, 1997.[2]

OTAG ASSESSMENT RESULTS

OTAG analytical work began with the development of an improved emissions inventory from which extensive modeling of ozone transport could be conducted. Based on that modeling, two workgroups reached several conclusions with respect to ozone transport and the anticipated benefits from increased nitrogen oxide (NO_x) and volatile organic compound (VOC) control. Among those findings particularly relevant in light of EPA's new 8-hour ozone National Ambient Air Quality Standard (NAAQS)[3] are the following:

The perceived contribution of ozone transport is strongly dependent on how the ozone "problem" is defined. Local emissions are more important with respect to peak 1-hour

[*] Excerpted from CRS Report 98-236. Updated June 15, 1999.

[1] Nichols, Mary D. "Ozone Attainment Demonstration," U.S. EPA: Assistant Administrator for Air and Radiation, March 2, 1995.

[2] Background on OTAG and its final report are available via EPA's web site at *http://www.epa.gov/ttn/otag/*

[3] For a summary of the new ozone NAAQS, see CRS Report 97-721 ENR, *Air Quality Standards: EPA's Final Ozone and Particulate Matter Standards*. For a longer discussion, see CRS Report 97-8 ENR, *Air Quality: EPA's New Ozone and Particulate Matter Standards.*

concentrations than with respect to lower concentration thresholds and concentrations assessed over longer averaging times (8-hour or seasonal averages), where larger areas and longer distance scales become increasingly important.

The central portion of the [37-State] OTAG domain is unique with respect to ozone and ozone transport. It persistently has elevated ozone levels producing an "ozone pool." Transport in any direction from this region has been implicated with high ozone levels in neighboring areas.

High ozone levels in the southern portion of the OTAG domain are typically associated with stagnant [pollutant] transport conditions resulting in shorter transport scales than on average. In contrast, high ozone levels in the northern portion of the OTAG domain are more typically associated with higher speed and persistent transport conditions from inside the OTAG domain across state lines.

The proposed 8-hour ozone standard [now finalized] will result in significantly more closely adjacent nonattainment areas across the OTAG domain. This will make ozone transport more critical with respect to nonattainment than it is under the current [now previous] standard.[4]

OTAG Recommendations[5]

Of the 37 states that constituted OTAG, 32 voted in favor of the final recommendations with Alabama, Kentucky, Michigan, Virginia, and West Virginia voting against. Those recommendations covered numerous areas, ranging from utility controls to increased research.

Size of Transport Region

Which states should comprise the ozone transport region was debated continuously during OTAG deliberations. Particularly, states west of the Mississippi River and some southern states argued that their emissions had little or no effect on ozone transport. OTAG recommended that these "coarse-grid" states be exempt from OTAG-related controls. These states include North Dakota, South Dakota, Nebraska, Kansas, Oklahoma, Texas, Minnesota, Iowa, Arkansas, Louisiana, Mississippi, and Florida, as well as parts of Maine, New Hampshire, Vermont, New York, Michigan, Wisconsin, Missouri, Alabama, and Georgia.

OTAG did recommend that Iowa work with Wisconsin in developing implementation plans for southwestern Wisconsin; that Kansas work with Missouri on implementation plans for Kansas City; and, that Oklahoma, Texas, Arkansas, and Louisiana share modeling results with Missouri.

[4] Air Quality Analysis Workgroup. *OTAG Air Quality Analysis Workgroup Results Summary. http://capita.wustl.edu/otag/Reports/ExecSumm/AQASUM3.html*

[5] OTAG's recommendations are attached to EPA's proposed Ozone Transport Rule as Appendix B. 62 *Federal Register* 60376-60379, November 7, 1997.

Utility NO_x Controls

Increasing NOx control on utility plants was one of the most contentious issues within OTAG. The final recommendation reflects this contention by being wide enough to cover all points of view. Specifically, OTAG recommended a range in the 26 "fine grid" states between current Clean Air Act controls (i.e., existing requirements) to the lesser of 85% reduction from 1990 rates or a .15 lb/mmBtu performance standard (a significant increase in stringency). It is arguable how such a wide range provided any guidance to EPA on implementing regional controls. Essentially, the range gave EPA a free hand in deciding on appropriate tonnage levels in its proposed Ozone Transport Rule, discussed later.

OTAG did recommend that any statewide tonnage levels proposed by EPA be revised if additional subregional modeling or air quality analyses justified it. Likewise, OTAG recommended that EPA consider any submitted comments and subregional modeling with respect to proposed statewide tonnage levels prior to finalizing SIP calls.

Non-Utility Stationary Source NOx Control

OTAG's recommendations for non-utility stationary sources mirrored those for the utility sources. Dividing non-utility facilities into medium and large sources, OTAG recommended that large sources be controlled in proportion to how much utility sources are controlled. If utility sources are required to reduce 55%, large non-utility sources should be required to reduce 55%. This parallel requirement weakens as the utility reductions increase; for an 85% utility source reduction, OTAG recommended only a 70% reduction for large non-utility sources.

Requirements for medium sources are also based on utility source requirement, but considerably less stringent. For 55%-65% utility reductions, no medium non-utility control is recommended. For 75%-85% utility reduction, OTAG recommended that medium non-utility sources be required to install reasonably available control technology (RACT), a determination made by the states based on EPA guidance. As with the utility control recommendations, OTAG urged flexibility if additional modeling and air analyses warrant it.

Federal Emission Control Measures

In OTAG's view, the federal government has a major role in assisting states achieve compliance, particularly with respect to new stationary source control and mobile source control. OTAG outlined nine specific areas where the federal government should develop, adopt, and implement stringent national NO_x and/or VOC control requirements to assist the OTAG region in achieving compliance with the ozone NAAQS. These areas include architectural and industrial maintenance coatings, consumer and commercial products, autobody refinishing, reformulated gasoline, small engines, heavy duty highway engines, heavy duty non-road diesel engines, and locomotive engines.

OTAG also recommended that EPA act on some fuel-related items. These recommendations included EPA (1) adopting a rule on an appropriate fuel sulfur standard to improve long-term performance of emission control devices, (2) evaluating emission benefits and other effects of cetane adjustments on current diesel engines and adopting standards as appropriate, and (3) considering and potentially adopting new diesel fuel standards by the year 2004 through existing deliberative processes.

Finally, OTAG supported and encouraged adoption of a National Low Emission Vehicle Program.

Specific State Actions

OTAG also recommended that "fine grid" states consider adopting enhanced inspection and maintenance (I&M) program in urban areas with populations greater than 500,000, the continued use of reformulated gasoline (RFG) in both mandated and opt-in areas, among other provisions.

Trading Program

Part of the controversy over utility source control involved proposed market-based approaches to implementing such controls. OTAG did considerable work on market mechanisms and found several benefits to these implementation approaches, including: (1) lower compliance costs, (2) incentive for early reductions, (3) incentive for over-control of sources, (4) incentive for innovation, and (5) administrative flexibility.

However, OTAG was unable to agree on a basic structure for a market mechanism. Some states, particularly those in the northeast, advocated a "cap and trade" program. A cap and trade program places a total emission limit (i.e., cap), expressed in tons of pollutants, on specific emission sources within an area. The allowable amount of pollution is then allocated to these sources according to an agreed upon formula. Sources that emit less pollution than their allocation may trade the unused portion of their allocation to sources that exceed their allocation, or bank those unused pollution "credits" for use in a future year. A system like this has been successfully implemented to reduce sulfur dioxide (SO2) emissions under Title IV of the CAA and is being implemented in the Ozone Transport Region (OTR) to reduce NO_x emissions.

Under a cap and trade system, emissions are not permitted to increase over time. Thus, new emissions from new sources (or expanded production at existing sources) have to be offset with reduced emissions at existing sources. Industry groups have objected to this fixed limit on emissions and argued for an alternative that controls the rate of pollution emitted, but not the total amount emitted (i.e., no cap). Under a "rate only" approach, a source could emit as much of a pollutant as it wished, as long as it did not emit the pollutant at a rate greater than stipulated. If a source emitted at below the stipulated rate, it would receive credits amounting to the difference between its actual rate and the stipulated rate times its production for a given year. It could trade or bank the resulting credits. If a source exceeded the rate, it could purchase emission credits for the excess amount (based on its total production for a year) from a source that had accumulated credits. Currently, no trading program is based on this approach.

As was the case with the utility control recommendations, OTAG's final recommendation with respect to a trading program encompassed both views. The cap and trade program, called

"Track 1," and the rate only program, called "Track 2," were both identified by OTAG as options for states to adopt in implementing utility (and possibly other) control measures. OTAG acknowledged several uncertainties with respect to this recommendation including (1) the need for further work on rate only approaches to implement them with the same level of confidence and certainty as the cap and trade system; and (2) how any "cross-track" trading could occur between cap and trade states and rate only states. OTAG did note that EPA would have an integral part in overseeing the implementation and integrity of these approaches.

For its part, OTAG recommended a joint state/EPA workgroup address the issues of implementation of Track 1 and Track 2, along with identifying key design features that states could select in developing a NO_x trading program. OTAG believed its work in this area would provide a sound basis for the workgroup's task. In addition OTAG suggested the workgroup address other concerns with respect to modeling, market systems, and local control requirements.

EPA's Revised NAAQS for Ozone

On July 16, 1997, the Administrator of the Environmental Protection Agency (EPA) finalized a new National Ambient Air Quality Standard (NAAQS) for ozone. Accompanying the NAAQS was a White House memorandum discussing the general approach EPA was to take in implementing the new NAAQS.[6] Recognizing ozone as a regional pollution problem, the memorandum focused on regional compliance strategies for the new NAAQS. In particular, the memorandum cited the work of OTAG as the basis of this approach. As stated therein:

> For the past 2 years the EPA has been working with the 37 most eastern states through the Ozone Transport Assessment Group (OTAG) in the belief that reducing interstate pollution will help all areas in the OTAG region attain the NAAQS.... The OTAG completed its work in June 1997 and forwarded recommendations to the EPA. Based on these recommendations, in September 1997, the EPA will propose a rule [on ozone transport] requiring states in the OTAG region that are significantly contributing to nonattainment or interfering with maintenance of attainment in downwind states to submit SIPs [State Implementation Plans] to reduce their interstate pollution. The EPA will issue the final rule by September 1998.

The revised ozone NAAQS was not the trigger for the Ozone Transport Rule finalized in October 1998, and discussed in the next section. Existing nonattainment in the Northeast corridor with the previous, less stringent ozone NAAQS was the genesis of the OTAG effort and the resulting rule by EPA. However, EPA drew on the OTAG findings and recommendations as it developed a flexible implementation strategy for attainment of the new ozone NAAQS, and, where EPA considered it appropriate, this strategy is integrated with the requirements of the Ozone Transport Rule. In particular, the regional NO_x control strategy advanced by EPA in the Ozone Transport Rule as a response to ozone transport under section 110(k)(5) is also an integral part of the new ozone NAAQS implementation strategy under section 110(a)(1)). In particular, EPA believes that implementation of a regional NO_x control strategy will permit many areas that are (or would be by the year 2000) in attainment with the

6 President Clinton. "Implementation of Revised Air Quality Standards for Ozone and Particulate Matter," Memorandum for the Administrator of the Environmental Protection Agency, July 16, 1997.

previous ozone NAAQS but in non-compliance with the new ozone NAAQS, to come into compliance with the new standard with little or no additional new local emission reductions. EPA says that participation by these "transitional areas" in the regional NO_x control program would permit EPA to eliminate unnecessary local planning requirements for such areas, and to revise its new source review (NSR) and conformity rules so such transitional areas could comply with only minor revisions to their existing programs.

OZONE TRANSPORT RULE

On October 27, 1998, the EPA finalized its Ozone Transport Rule.[7] The rule requires 22 eastern states and the District of Columbia to submit state implementation plans (SIPs) to address regional transport of ozone.[8] In particular, the rule calls for increased controls on NOx, focusing particularly on emissions from electric utilities and large combustion sources. To achieve the necessary reductions, EPA has stipulated emission budgets for each of the affected states, with each state free to decide on what controls to use to maintain emissions within those budgets. As discussed in the next section, EPA is also encouraging the formation of a regional cap and trade program to implement the NO_x reductions through a model program.[9] As finalized, the rule requires the affected states to submit SIP revisions to EPA by September 1999 and to implement controls to achieve their state emission budgets by May 1, 2003. By September 30, 2007, states are expected to be in compliance with their NO_x budgets.

Under section 110(k) of the Clean Air Act, EPA is mandated to require SIP revisions for states whose current SIPs are substantially inadequate to mitigate adequately interstate pollutant transport such as that described under section 176A (under which OTAG was established). Adequate provisions are defined by section 110(a)(2)(D) as those that prohibit state emissions "which will contribute significantly to nonattainment in, or interfere with maintenance by, any other state with respect to any such national primary or secondary ambient air quality standard...." In the final rule, EPA identified four criteria for determining whether an upwind state's contribution to a downwind state's nonattainment status was "significant."[10] Three of the four criteria identified by EPA related to air quality: (1) collective contribution of regional emissions to the problem; (2) extent of downwind nonattainment problems, including ambient impact of required controls; and, (3) ambient

[7] Environmental Protection Agency. Finding of Significant Contribution and Rulemaking for Certain States in the Ozone Transport Assessment Group Region for Purposes of Reducing Regional Transport of Ozone. Rule. 63 Federal Register 57356-57538, October 27, 1998. For the rule as proposed, see: Environmental Protection Agency. Finding of Significant Contribution and Rulemaking for Certain States in the Ozone Transport Assessment Group Region for Purposes of Reducing Regional Transport of Ozone. Notice of proposed rulemaking. 62 Federal Register 60317-60421, November 7, 1997

[8] The 22 states included are: Alabama, Connecticut, Delaware, Georgia, Illinois, Indiana, Kentucky, Massachusetts, Maryland, Michigan, Missouri, North Carolina, New Jersey, New York, Ohio, Pennsylvania, Rhode Island, South Carolina, Tennessee, Virginia, Wisconsin, and West Virginia.

[9] On April 29, 1998, EPA proposed a supplemental rulemaking that detailed a model cap and trade program for states to consider in implementing the Ozone Transport Rule. Those details are incorporated in the final rule. Environmental Protection Agency. Supplemental Notice for the Finding of Significant Contribution and Rulemaking for Certain States in the Ozone Transport Assessment Group Region for Purposes of Reducing Regional Transport of Ozone; Proposed Rule. 63 Federal Register 25902-25994, May 11, 1998.

[10] These were modifications of four criteria listed in its November 1997 proposal: (1) the emissions level of the upwind area; (2) the upwind area's contribution to the downwind nonattainment area; (3) the transport distance

impact of emissions from upwind sources. The fourth criterion was economic — the availability of cost effective control measures for upwind emissions.[11]

Applying the three air quality related criteria to findings from the OTAG modeling process and other modeling efforts, EPA determined that the "weight of evidence" indicated that 22 of OTAG's 37 member states made significant contributions to downwind nonattainment--the 22 plus D.C. that are addressed by the rule.[12] For the other 15 states, EPA took no action in the final rule, and is not requiring that they submit SIP revisions at this time.[13] This "no action" determination by EPA is a change from its proposed rule that found that these states did not meet EPA's weight of evidence threshold for finding a contribution significant.[14]

As originally proposed, EPA used OTAG and other modeling efforts to determine state complicity with respect to ozone transport, but did not use such modeling to determine its proposed emissions budgets for affected states.[15] Instead, EPA used a cost-effectiveness criterion based on NO_x control costs to determine allocations. Budget components were calculated for five sectors: electric utility sources, nonutility stationary sources, area sources, nonroad engines, and highway vehicles.[16] In focusing on cost-effective controls, EPA argued that it was following the recommendations of OTAG. The effect of such an allocation scheme would be to tilt emission reductions toward states with large stationary NO_x sources with relatively modest controls, and away from states with relatively higher emissions from other sources. In particular, states with coal-fired electric power plants would generally have higher percentage reductions required on a statewide basis than states without such plants. This focus on regional reductions in NO_x emissions, rather than attempting to target reductions based on air modeling, was further defended by EPA in the final rule. As stated by EPA, EPA:

between the two areas; and (4) the geographic extent of the contribution downwind. 62 Federal Register 60335, November 7, 1997.

[11] 63 Federal Register 57376-7, October 27, 1998.

[12] The 23 jurisdiction were found to"make a significant contribution to nonattainment of the 1-hour and 8-hour NAAQS in, or interfere with maintenance of the 8-hour NAAQS..." 63 Federal Register 57398, October 27, 1998

[13] The 15 states not included are: Arkansas, Florida, Iowa, Kansas, Louisiana, Maine, Minnesota, Mississippi, Nebraska, North Dakota, New Hampshire, Oklahoma, South Dakota, Texas, and Vermont.

[14] In the proposed rule, EPA found that the 15 states: (1) made at most a relatively small contribution to downwind nonattainment areas; (2) most are relatively distant from many of the downwind nonattainment areas; and (3) most have a relatively low amount of manmade NO_x emission and/or NO_x emissions density. Thus, these states did not meet EPA's weight of evidence threshold for finding a significant contribution.

[15] "The statewide emissions budgets proposed in this rulemaking were not modeled directly to determine their air quality benefits." 62 Federal Register 60327. In its April proposed supplemental rulemaking, EPA included a modeling analysis to show how these proposed NO_x budgets would improve air quality in the eastern U.S. See 63 Federal Register 25953-25967, May 11, 1998.

[16] For utility sources, EPA used a NO_x emission rate of 0.15 lb/MMBtu to determine budget allocations. However, EPA used different growth factors than those developed by OTAG in making the necessary 2007 calculations. For area sources, EPA assumed no new controls, in line with OTAG. For nonutility sources, EPA used a 70% reduction requirement for large and RACT controls (generally 25%-50% reduction) for smaller sources. This is approximately what OTAG had recommended. EPA calculated the highway vehicle budget by assuming implementation of existing SIPs, along with the following federal measures: national low emission vehicle standards, 2004 heavy-duty engine standards, and revisions to emissions test procedures. EPA's method of calculating budgets for this sector differs somewhat from OTAG's recommendations. Finally, EPA calculated budgets for nonroad engines assuming implementation of existing SIPs, along with the following federal measures: federal small engine standards (Phase II), federal marine Engine standards (diesels >50 hp), federal locomotive standards, and 1997 proposed nonroad diesel engine standards. EPA used OTAG growth projections in calculating the budgets for this component. See proposed Ozone Transport Rule and Appendix B (OTAG Recommendations), 62 Federal Register 60318-60420, November 7, 1997.

> ...acknowledges air quality modeling makes clear that reductions in emissions closer to the air quality problem have a greater ambient impact. However, EPA has not been presented with, nor been able to develop, an accurate comparison of the downwind costs of emission reductions that would achieve the same ambient impact as the regional reductions required by today's action. The EPA does not have comprehensive information concerning available local measure or their costs or ambient impacts.
>
> However, as a qualitative matter, EPA believes that available evidence indicates that the upwind costs are reasonable not only in light of cost-effectiveness per ton removed, but also in light of the downwind ambient impact of the emissions reductions. Under the 1-hour NAAQS, emissions from each upwind State generally affect several downwind nonattainment urban areas. Thus, matching the total ambient impact of the emissions reductions from the upwind State would require emission reductions in several downwind areas. [footnote omitted] Although presently available information does not permit a useful quantitative comparison of total upwind and downwind costs in terms of their ambient impact, EPA believes that upwind reductions replace local reductions that, on a cost-per-ton removed basis, may be expected to be more expensive.[17]

As a final argument for its budget allocations based on regional cost-effectiveness criteria, EPA states that modeling various regional variations in upwind emission reductions indicated no subregional variations in upwind emission reductions that were either more cost-effective or more beneficial to air quality than the approach it chose to take.

The budget allocations resulting from this approach were revised by EPA in its final rule from the revisions contained in its April 1998 proposed supplemental rulemaking.[18] At first glance, it would appear that the 1.16 million ton (28%) reduction under the final rule is less stringent than the 1.59 million tons (35%) originally proposed. However, the apparent lessening of the reduction requirement is primarily due to downward revisions in the base inventory, not because of an increase in permitted emissions. In particular, the final rule lowers the base case to account for various mobile source controls anticipated to occur by year 2007, and to account for EPA's decision to remove more than 11,000 smaller industrial boilers from its budget calculations. The actual difference between the budget proposed and the budget as finalized is only 77,000 tons.

The budget allocations were revised again by EPA in a technical amendment issued in May 1999.[19] These revisions result mostly from a recent settlement on NO_x emissions from heavy-duty diesel engines. The changes lower the anticipated reduction from 1.16 million tons (28%) under the final rule to 1.14 million tons (25%). The amended final emissions budget for each state is provided in Table 1. All the states with reduction requirements at 30% or above are from the Midwest. In contrast, states in the Northeast have budget allocations generally well below the nationwide average of 25%. This disparity reflects EPA's focus on electric utilities as the most cost effective source of NO_x reductions. Of the 1.14 million ton reduction estimated by EPA to occur under the rule, 0.96 million is estimated to come from electric utilities. As a whole, this would be a 64% reduction from EPA's baseline for electric utilities emissions. For some Midwestern states, the reduction is over 70%.

---

[17] 63 Federal Register 57405, October 27, 1998.

[18] See 63 Federal Register 25904-25911 (May 11, 1998) for further detail on revisions.

[19] See 63 *Federal Register* 25918-25950, May 11, 1998.

Table 1. Seasonal NO_x Emissions Budget for States Making a Significant Contribution to Downwind Ozone Nonattainment (Tons of NO_x per Ozone Season)

State	2007 Base Emissions	Proposed 2007 Budget	% reduction
Alabama	236,867	172,037	25%
Connecticut	46,220	43,081	7%
Delaware	23,512	22,789	3%
District of Columbia	6,485	6,672	-3%
Georgia	253,489	189,634	25%
Illinois	375,250	274,799	27%
Indiana	355,433	238,970	33%
Kentucky	238,412	155,619	35%
Maryland	103,558	81,625	21%
Massachusetts	87,563	85,298	3%
Michigan	288,000	224,582	22%
Missouri	189,737	128,146	32%
New Jersey	108,584	100,133	8%
New York	253,659	240,123	5%
North Carolina	228,600	168,373	26%
Ohio	378,418	250,930	34%
Pennsylvania	346,900	257,441	26%
Rhode Island	9,895	9,810	1%
South Carolina	153,465	124,211	19%
Tennessee	257,962	197,664	23%
Virginia	224,521	185,027	18%
West Virginia	184,947	91,216	51%
Wisconsin	175,061	136,172	22%
Total	**4,526,538**	**3,384,350**	**25%**

Source: U.S. EPA.

In addition to the emission budget identified above, the rule contains an additional feature that could affect a state's budget in specific circumstances. Because of concerns raised about the possible effects of the rule on electricity reliability, EPA established a compliance supplement pool of 200,000 tons that are allocated to the individual states in proportion to the size of reduction they must achieve under the rule. The states may use their pool to allocate emission tons to sources that reduce emissions prior to May 2003 (early reductions) or to sources that demonstrate need for relief from the compliance deadline. For this latter allocation, sources requesting such treatment would have to demonstrate that the compliance deadline creates "undue risk" for the source or its associated industry (e.g., the electric utility industry). Any tons not distributed by a state prior to May 1, 2003 will be retired by EPA.

MODEL NO_X CAP AND TRADE PROGRAM

As expected, EPA intends to work with states to implement the above NO_X budgets through an interstate cap and trade program. The details of that program were first provided in the supplemental notice of proposed rulemaking issued in April 1998 and refined in the final rule.[20]

Following the example set by the acid rain trading program, the EPA model NO_X program is an allowance-based cap and trade program (OTAG's "Track 1" recommendation). One allowance equals one ton of NOx, with each participating state receiving allowances equal to the emission budgets of those sources participating in the program.[21] Unlike the acid rain program where participation and individual allowance allocations are mandated by federal law, state participation in the NO_X program is voluntary, and the allowance allocations to the individual emissions sources is determined by the state, not the federal government. The states also approve monitoring systems based on federal regulations and enforce compliance. Beyond approving and overseeing the SIPs, EPA's role is primarily administrative in terms of collecting data, allocating allowances as prescribed by states, recording trades, and reconciling emissions and allowance data every year. In particular, EPA will track allowances to facilitate interstate trading and enforcement.

Although EPA's model rule permits states to determine individual source allowance allocations, it requires certain elements of the individual state programs to be consistent in order for them to participate in interstate transactions. These elements include the timing of allowance allocations, monitoring requirements, and noncompliance penalties. In particular, the program must be a tonnage cap-based program (i.e., no rate-based program — OTAG's "Track 2" recommendation) to be included in the regional trading system. Indeed, the SIP approvability criteria virtually ensure that, with respect to large combustion sources, states will have to submit a tonnage based program to get the necessary approval from EPA.[22] Besides the cap requirement, EPA requires states, if they choose to participate in the program, to submit allowance allocation for participating sources by September 30, 1999 for the control periods in the years 2003, 2004, and 2005. Afterwards, states must submit allocations annually for the control period three years hence. Such allocations may not exceed the state's emission budget. At a minimum, the program must include any electric generating unit greater than 25 Mw that sells power and any steam unit with a maximum design heat input greater than 250 mmBtu/hr (such as industrial boilers). In addition, the state must agree to the

---

[20] EPA will initially require that participating sources include existing fossil fuel fired electric generating plants greater than 25 megawatts and other existing boilers and turbines with a heat input greater than 250 mmBtu/hr. These core sources represent 80% of the 2007 baseline emissions inventory, 65% of the 2007 emissions cap inventory, and 90% of the required reductions under the proposed rule. Participation by sources outside these core groups would be determined by the individual states. 63 *Federal Register* 57518, October 27, 1998.

[21] As explained in the proposed notice of supplemental rulemaking, states may choose one of three ways for requiring large combustion sources to meet their limitations under the transport rule: (1) a tonnage limit for each source that adds up to the allowed budget; (2) an emission rate limit that when multiplied by the maximum capacity for each source adds up to the allowed budget; and, (3) using alternative method providing equivalent assurance that the allowed budget will not be exceeded. 63 *Federal Register* 25912, May 11, 1998

[22] The model rule includes banking of unused allowance in a given year to use in a future year. However, there is a "flow" restriction on the use of banked allowances — the total number of banked allowances used in a given year may not exceed 10% of a state's emissions budget. Any banked allowance used above that amount is subject to a 2-for-1 penalty (must use 2 allowances to offset 1 ton of emissions). The model rule's monitoring and tracking system is based on the current system used for the SO_2 program under Title IV. See 63 *Federal Register* 57529-57535, October 27, 1998.

emission banking and monitoring provisions contained in the model rule.[23] States choosing to use the federally sponsored program will receive "streamlined" approval from EPA.

States have some flexibility with respect to the model rule in terms of who they include, the allocation scheme employed, and how they determine any early reduction credits. States may expand the applicability of the trading program to smaller sources or other sources not included in the model rule. Although EPA proposes an allocation scheme to states, states are free to allocate allowance in any manner that seems appropriate to them. Likewise, states have flexibility in determining early reduction credits. However, if a state chooses to exercise these opportunities, the approval of the resulting SIP by EPA may be slowed.

Compared with a traditional command and control implementation strategy, EPA originally projected a regional cap and trade program that would reduce compliance costs to the electric utility industry by about one-fourth — from about \$2.1 billion to \$1.6 billion annually. In its proposed supplemental rulemaking, EPA revised the cost estimate for electric utilities from \$1.6 billion to \$1.4 billion annually. In addition, EPA estimated that a regional cap and trade program would reduce compliance costs to non-electric generating sources from \$1.2 billion to \$456 million annually.[24] Comparative estimates for other sectors were not available. In the final rule, EPA refers to the Regulatory Impact Assessment (RIA) conducted for the rule with costs estimates of \$1.7 billion annually (1990 dollars) and quantifiable benefits of between \$1.1 billion and \$4.2 billion.[25]

PROPOSED FEDERAL IMPLEMENTATION PLANS (FIPS)

In conjunction with its final rule on the ozone transport SIP call, EPA proposed a Federal Implementation Plan (FIP) for any state that fails to respond adequately to the SIP call.[26] Basically, the purpose of a FIP is to correct, supplement, or replace a SIP determined by EPA to be inadequate.[27] Under a FIP, EPA may exercise the authority granted to the state under the Clean Air Act. The EPA is required to promulgate a FIP at any time within 2 years of finding a state SIP submittal either inadequate or non-existent, or disapproving a SIP submittal in whole or in part — unless the state corrects the deficiency to EPA's satisfaction before such promulgation.

Historically, EPA has been reluctant to impose FIPs on states that submit inadequate SIPs or no SIP at all. This is not the case with the ozone transport rule. Under the final ozone transport rule, states are required to submit SIPs by September 1999. With this proposed FIP rule, EPA would initiate a FIP for each of the 23 jurisdictions covered by the SIP call. Thus, EPA would be able to move quickly with a FIP if a state fails to submit a SIP by September 1999, or if subsequent to the deadline, EPA finds a SIP submittal inadequate. The timing of this FIP proposal suggests that EPA intends to aggressively assert its authority under the CAA to ensure compliance with the new ozone transport rule. As stated by EPA in the proposed rule:

---

[23] 63 *Federal Register* 25911, May 11, 1998. Estimates are in 1990 dollars.

[24] Environmental Protection Agency. *Regulatory Impact Analysis for the Regional NO_x SIP Call.* Office of Air Quality Planning and Standards, September 1998.

[25] Environmental Protection Agency. *Federal Implementation Plans to Reduce the Regional Transport of Ozone.* Proposed Rule. 63 *Federal Register* 56394-56427, October 21, 1998.

[26] See CAA, section 110(c)(1).

[27] 63 *Federal Register* 56399, October 21, 1998.

> Thus, although the CAA allows EPA up to 2 years to promulgate a FIP after a finding of a State's failure to submit a complete, approvable plan, EPA intends to expedite the FIP promulgation to help assure that the downwind States realize the air quality benefits of regional NO_x reductions as soon as practicable.

Another possible reason for EPA's aggressive attitude with respect to FIP may be the model trading scheme EPA is hoping states will choose to implement the ozone transport rule. As noted previously, under the SIP process EPA does not have the authority to impose a trading scheme on the states as it does under Title IV. However, under the FIP process, EPA does. The implementation plan imposed under the proposed FIP rule is largely based on the model trading program suggested to states under the SIP call. In addition, the proposed FIP rule provides that all state trading schemes covered under it are identical, an important condition to ensure maximum efficiency in the operation of the scheme.[28] This federal allocation of allowances and federal accounting of trading brings the potential operation of the NO_x trading program closer to the system EPA developed under Title IV. Also, the proposed FIP rule may reinforce on the states the kind of trading program they would adopt under a SIP in order to be coordinated with any FIP program, along with ensuring easier EPA approval of their SIP. As hinted by EPA: "EPA does not have the statutory authority under Title I to promulgate a Federal cap-and-trade program to achieve a State's SIP call budget *unless the State fails to respond adequately to the SIP call* [emphasis added]."[29]

Substantively, the implementation scheme under the proposed FIP rule covers electric utilities and large non-utility stationary sources (including large industrial boilers, industrial turbines, internal combustion engines, and cement manufacturing). The electric utilities and large industrial sources (boilers, turbines and combined cycle) would be covered under a mandatory federal emission trading program. The FIP would use the same source cutoff levels, categories, and control levels as used to develop the SIP call's emission budget, and would require compliance by the same May 1, 2003 deadline.[30]

SECTION 126 PETITIONS

On May 25, 1999, the EPA finalized its finding with respect to petitions submitted by eight northeastern states.[31] Submitted in August 1997 by Connecticut, Maine, Massachusetts, New Hampshire, New York, Rhode Island, Pennsylvania, and Vermont, the petitioners asked EPA to make a finding that utilities and/or major sources of NO_x in the Midwest contribute

---

[28] 63 *Federal Register* 57458, October 21, 1998.

[29] 63 *Federal Register* 56401, October 21, 1998.

[30] All the petitions targeted the Midwest; however, some included other sources in the South, Southeast, and Northeast. A discussion of the petitions are provided in the Ozone Transport Rule. A copy of the Memorandum of Agreement Concerning Schedule for EPA Action on Section 126 Petitions is available from EPA's web site: http://www.epa.gov/airlinks/

[31] Midwestern states, which would bear the brunt of any reductions imposed if EPA takes action under the Section 126 petitions, have objected to the agreement. They claim that the northeastern states have not themselves taken adequate actions to address ozone pollution, that reductions imposed on the Midwest would not solve the petitioning states' problems, and that the costs to the Midwestern states in reducing emissions would not be commensurate with benefits. An attempt to block EPA from acting on the petitions was filed by Midwestern utilities in the U.S. Court of Appeals for the D.C. Circuit. This lawsuit was dismissed by the court on April 28, 1998, since EPA's actions "do not constitute final appealable orders." (*Midwest Ozone Group v. EPA*, D.C. Cir., No. 97-1627, 4/28/98)

significantly to the ozone nonattainment problem in their states.[32] All eight states directed their petitions to the statutory 1-hour ozone standard; only three states — Massachusetts, Pennsylvania, and Vermont — also directed their petitions to the new 8-hour standard. In November 1998, two of the other states — Maine and New Hampshire — requested EPA to evaluate their petitions under the 8-hour standard in addition to the 1-hour standard. In April 1999, New Jersey and Maryland became the ninth and tenth states to submit a Section 126 petition to EPA seeking relief from out-of-state sources of ozone precursors. EPA has 60 days to respond to these new petitions (with a 6-month extension possible, if EPA determines it necessary).

As noted earlier, if EPA found any of the petitions technically meritorious, it would have to establish federal emission limits for the sources affected by the petitions. In its final rule, EPA determined that portions of six of the eight petitions initially submitted were technically meritorious under Section 126 and deemed granted at future dates depending on how affected states respond to the ozone transport rule discussed above.[33] Determinations were not made on the two recently submitted petitions from New Jersey and Maryland.[34]

The final rule finds portions of six of the eight petitions technically meritorious under Section 126 under the 8-hour standard, and four petitions (Connecticut, Massachusetts, New York, and Pennsylvania) under the 1-hour standard. Based on the same methodology and data sources used for the Ozone Transport Rule, EPA found 19 states and the District of Columbia as significantly contributing to the nonattainment of either the 1-hour or 8-hour ozone NAAQS, or in the maintenance of attainment within at least one of the petitioning states. The 20 entities are: Alabama, Connecticut, Delaware, District of Columbia, Illinois, Indiana, Kentucky, Maryland, Massachusetts, Michigan, Missouri, New Jersey, New York, North Carolina, Ohio, Pennsylvania, Rhode Island, Tennessee, Virginia, and West Virginia.[35]

To implement the finding, EPA's rule rests heavily on the analysis and trading program contained in the SIP call. Adjustments to the SIP call analysis were necessary because of constraints rendered by Section 126; for example, Section 126 only affects stationary sources. In addition, the removal of three states from the reduction pool meant reallocation of state budgets. Administration of the program would be similar to EPA's proposed FIP rule. Allocation of emission budgets to individual sources will be based on a 0.15 lb/mmBtu emission rate for electric generating units over 25Mw, and a 60 percentage reduction requirement for industrial boilers and turbines greater than 250 mmBtu/hr. No controls are required for smaller sources. As with the proposed FIP rule (but unlike the SIP call), participation in the trading program would be mandatory. Banking provisions, a set-aside program for new sources, and compliance supplement pools would also be included as

[32] On February 25, 1998, the eight petitioning Northeast states asked a federal court to make the agreement and schedule binding (*Connecticut v. Browner*, D.C. SNY, No 98CIV.1376, 2/25/98).

[33] Environmental Protection Agency. *Findings of Significant Contribution and Rulemaking on Section 126 Petitions for Purposes of Reducing Interstate Ozone Transport.* Advance Notice of Proposed Rulemaking. 63 *Federal Register* 24058-24061; EPA. *Findings of Significant Contribution and Rulemaking on Section 126 Petitions for Purposes of Reducing Interstate Ozone Transport.* Proposed Rule. 63 *Federal Register*, 56292-56391, October 21, 1998

[34] The petition from Vermont would be denied under the proposed rule because Vermont does not have, nor is it projected to have, any areas out of attainment with either the 1-hour Ozone NAAQS, or the new 8-hour Ozone NAAQS.

[35] Three states — Georgia, South Carolina, Wisconsin — are covered under the SIP call but not the proposed Section 126 finding because modeling indicates that they did not have a significant effect on nonattainment in the seven petitioning states.

mandatory provisions under EPA's implementation plan. Details on the source NO_x allocations and trading program are to be finalized by EPA in July 1999

IMPLEMENTATION SCHEDULES AND CHALLENGES

The parallel actions of the Section 126 petitions and the Section 110 SIP call presents difficulties for EPA as the two provisions are implemented differently. The Ozone Transport Rule is a SIP call under section 110(k)(5) for revised SIP provisions meeting the requirements of section 110(a)(2)(D) for the statutory 1-hour ozone NAAQS. Because of the new ozone NAAQS, this SIP call is coupled with a proposal under section 110(a)(1) for SIP submissions to meet the requirements of section 110(a)(2)(D) for the new 8-hour ozone NAAQS. Because EPA has found that individual states contribute significantly to ozone nonattainment downwind, it requires those states to submit SIP provisions to limit NO_x emissions to levels specified by EPA; the specific means and sources affected will be determined by the state. Only if a state does not make the required submission would EPA take enforcement actions, including promulgation of a federal implementation plan (FIP) for the state.

Section 126 does not work this way. Under Section 126, if a petition is granted, EPA is responsible for devising and implementing control strategies for the affected sources. Thus, unlike the SIP process where states determine control strategies to achieve mandated reductions, Section 126 requires EPA to select and impose appropriate controls on a state's affected emission sources. In effect, Section 126 implementation is more like a FIP than a SIP. Also, unlike the SIP process, which can coordinate strategy across all emitting sectors, Section 126 only addresses major stationary sources.

To reconcile these differences, EPA and the eight petitioning states entered into a Memorandum of Agreement for EPA action on the Section 126 petitions. The December 1997 agreement harmonizes the time frame for EPA action on the Section 126 petitions with EPA's schedule for action on the Ozone Transport Rule. In line with that schedule, EPA issued an Advance Notice of Proposed Rulemaking for the Section 126 petitions in April 1998, a proposed rule in October 1998, and a final rule in April, 1999. Highlights of the schedule as currently proposed by EPA are presented table 2.

Table 2. Proposed Schedule for Section 126 and Section 110 Actions

Date of Proposed Action	Proposed Action
April 1998	EPA publishes an advance notice of proposed rulemaking with respect to the Sec. 126 petitions [*accomplished*]
September 1998	(1) EPA publishes a notice of proposed rulemaking with respect to the Sec. 126 petitions; (2) EPA issues final Ozone Transport Rule establishing NO_x budgets [*accomplished*]
April 30, 1999	EPA takes final action on Sec. 126 petitions. Action includes determination of their merit and proposed remedy, but could postpone imposition of controls [*accomplished*]
September 1999	States submit SIP revisions to EPA in response to Ozone Transport Rule. EPA takes action to impose FIPs on states that fail to submit adequate plans
November 1999	If EPA chooses, EPA's April 1999 Sec. 126 action may be automatically imposed on affected states if EPA does not propose to approve those states' SIPs submitted under the Ozone Transport Rule
May 2000	If EPA chooses, EPA's April 1999 Sec. 126 action may be automatically imposed on affected states if EPA does not approve those states' SIPs submitted under the Ozone Transport Rule.
May 1, 2003	States implement SIP controls under the Ozone Transport Rule
September 30, 2007	States are expected to be in compliance with their NO_x emission budgets

Source: EPA, September 24, 1998

This attempt to harmonize and coordinate EPA action on the 1-hour and 8-hour standards and on the SIP call and the Section 126 findings has been challenged by a recent U.S. Court of Appeals decision on the 8-hour standard. Essentially, the Court ruled that deficiencies in EPA's promulgation of the 8-hour ozone standard (along with the particulate standards) required its return to EPA for further consideration. As a direct result of this decision, the Court has also suspended the SIP call, pending a full review. This decision, which will almost certainly be appealed by EPA, raises questions about EPA's compliance schedule and its attempt to harmonize its ozone transport rulings.

For the Ozone Transport Rule, a sustaining of the Court's decision on appeal could require EPA to revise the rule to focus exclusively on the 1-hour standard (which was the original trigger for the rule and was not challenged by the Court decision). Such a move seems unlikely to have any major effect on the substance of the Ozone Transport Rule. As indicated by Figure 1, the 22 states and the District of Columbia are affected by EPA's analysis of both the 1-hour standard and the 8-hour standard. If EPA were forced to revise the rule to affect only states that contributed to non-attainment of the 1-hour standards, the same 22 states (plus D.C.) would be affected. Obviously, the schedule for compliance might have to be revised to reflect the current suspension the rule, but the actual substance of the rule, in terms of who would have to reduce emissions, might not be much changed.

For the Section 126 petitions, the situation would be more complicated, if the decision is sustained on appeal. In many ways, the petitions submitted under Section 126 were designed to ensure the ozone reduction process in the eastern United States just in case the Ozone Transport Rule was stopped. As indicated previously by Table 2, if affected states do not respond to the transport rule by specified dates, EPA's findings with respect to the Section 126 petitions would kick in. With the suspension of the ozone transport rule, that situation becomes more possible.

This possibility is complicated by two factors. First, like the transport rule, EPA's findings under Section 126 couples nonattainment of the 1-hour standard with nonattainment of the 8-hour standard. However, unlike the ozone transport rule, the number of states affected under the 1-hour and the 8-hour standards differs. As indicated by Figure 2, while 19 states and the District of Columbia are affected under the 8-hour standard, only 12 states and the District of Columbia are affected under the 1-hour standard. If the decision is sustained on appeal, not only would states face uncertainty as to their mandate, but also the allocations may be affected.

A further complication is the existence and future possibility of more Section 126 petitions. As noted earlier, New Jersey and Maryland have already submitted petitions to EPA; petitions not reflected in the above discussion. If EPA's ozone transport analysis is any indication, the addition of New Jersey to the petitioning states would add Illinois to the states affected by the 1-hour standard. If other states follow New Jersey and Maryland, more additions are possible.

In short, the Court decision on the 8-hour standard may delay implementation of the Ozone Transport Rule and possibly EPA's findings with respect to the Section 126 petitions. The decision would appear to have little effect on the Ozone Transport Rule in terms of who is affected. However, depending on whether new Section 126 petition arise, the decision could result in fewer states being affected by EPA's Section 126 findings.

On June 11, 1999, EPA issued an interim final rule with respect to the Section 126 petitions, announcing its approach to the court decision. The rule temporarily stays until November 30, 1999, the effectiveness of the final rule on the Section 126 petitions published April 30, 1999. In response to the court decision on the 8-hour standard, EPA will decouple the Ozone Transport Rule from the Section 126 process and decouple implementation of the 1-hour standard from the challenged 8-hour standard. As a result, EPA will move forward with the Section 126 petitions based on the statutory 1-hour standard while it is appealing the court's decision with respect to the 8-hour standard and the resulting suspension of the transport rule. This would mean that Section 126 implementation would involve the 12 states and the District of Columbia as outlined above, at least until the situation with the 8-hour standard and the transport rule are clarified. EPA intends to take final action on these proposed changes to the Section 126 implementation process by November 30, 1999.

CONCLUSION

Ozone transport is a complex phenomenon; efforts to alleviate the effects of transport on downwind states will result in a mix of winners and losers, both in terms of states having to impose controls and in terms of states complying with ozone NAAQS. With major stakes involved, it is not surprising that some characterized OTAG's final recommendations as vague and ambiguous, where others saw needed flexibility. In any event, OTAG's work assessing ozone transport moved the debate along and clarified an issue made more important because of the new 8-hour ozone standard.

Overall, OTAG sanctioned flexibility, both for the states and for EPA in deciding on how to respond to the ozone transport issue and how to make necessary reductions. Although EPA often took the most stringent interpretation of OTAG's recommendations in its Ozone Transport Rule, it also has attempted to include flexibility for states in implementing proposed NO_x reductions. In particular, the sanctioning and encouragement of a regional NO_x cap and trade program for electric utilities and other major stationary sources could significantly reduce the cost of compliance with the rule. EPA does not have the authority to make participation in a regional trading program mandatory, unlike the current acid rain SO_2 program, unless a state fails to submit an adequate SIP and EPA proceeds to impose a FIP. Instead, it is up to the states to decide whether the benefits of reducing emissions across the enlarged trading area that the model trading program promises are worth the requirements EPA has imposed.

Implementation is also complicated by Section 126 petitions submitted by eight northeastern states. Implementation of controls under Section 126 is an EPA-directed affair, in contrast with the state strategies required under section 110. In addition, a Section 126 action could raise questions with respect to the Regulatory Flexibility Act as the controls imposed would clearly come from EPA and not the states (as is the case with section 110). In short, it is unclear whether the petitions under Section 126 will expedite NO_x control actions by affected states, or merely increase litigation on an already litigious issue. This litigiousness was most recently illustrated by the Court decision with respect to the 8-hour standard and suspension of the transport rule. More examples are likely as the rulemaking proceeds.

Chapter 9

AIR QUALITY AND VEHICLE EMISSION STANDARDS: AN OVERVIEW OF THE NATIONAL LOW EMISSION VEHICLE PROGRAM AND RELATED ISSUES*

David M. Bearden

INTRODUCTION

The Motor Vehicle Air Pollution Control Act of 1965 (P.L. 89-272) required the federal government to establish standards that limit the tailpipe emissions of each new vehicle manufactured for sale in the United States. While vehicle emission standards have become more stringent with subsequent amendments to the Clean Air Act in 1967, 1970, 1977, and 1990, the number of vehicle miles traveled has increased substantially from 588 billion in 1960 to over 1.5 trillion in 1995.[1] The rise in travel has offset a significant portion of the progress made in emission controls, and as a result, motor vehicles continue to be major sources of air pollution. The Environmental Protection Agency (EPA) reports that in 1996 on4oad vehicles accounted for roughly 60% of total emissions of carbon monoxide (CO) in the United States, about 31% of nitrogen oxides (NO_x), nearly 30% of volatile organic compounds (VOCs), and approximately 8% of particulate matter (PM).[2] NO_x and VOCs pose particular obstacles to improving air quality because they are the primary contributors to the formation of groundlevel ozone.[3]

Despite pollution controls currently in place, areas in many states have yet to attain the National Ambient Air Quality Standards (NAAQS).[4] Attainment of the standard for ozone

* Excerpted from CRS Report RL30005. January 4, 1999.

[1] Department of Transportation. Bureau of Transportation Statistics. FY1997 National Transportation Statistics. p.222.

[2] EPA. Office of Air Quality Planning and Standards. 1996 National Air Quality and Emissions Trends Report January 1998. p.82-86.

[3] Vehicles do not directly emit ozone. Rather, NO_x and VOCs produce ozone when they chemically react in the presence of sunlight. vehicles emit VOCs primarily in the form of hydrocarbons (HC).

[4] The NAAQs establish safe ambient levels for carbon monoxide, lead, nitrogen dioxide, ozone, particulate matter, and sulfur dioxide. States must develop plans for attaining the standards in areas that have exceeded allowable

has been a persisting problem especially among the Northeastern states primarily due to high concentrations of the ozone precursors NO_x and VOCs. While further reductions in emissions of these pollutants from motor vehicles might assist Northeastern and other states in complying with the ozone standard, the Clean Air Act prevents EPA from imposing more stringent emission standards on vehicle manufacturers sooner than model year (MY) 2004. To achieve additional emission reductions prior to this time, EPA has cooperated with numerous states and 23 vehicle manufacturers to develop a voluntary National Low Emission Vehicle (LEV) Program.[5] Manufacturers have agreed to produce cleaner vehicles for sale in eight Northeastern states and the District of Columbia beginning in MY1999 and nationwide in MY2OO1.[6] The program uses the same low emission standards and testing procedures already adopted in California, New York, Massachusetts, Vermont, and Maine. The main incentive for manufacturers to participate is that they will be able to produce a single fleet of vehicles for sale nationwide rather than having to produce a separate fleet to meet stricter requirements in only a few states.

While manufacturers voluntarily have agreed to produce cleaner vehicles nationwide, the National LEV program arose out of attempts by numerous states and EPA to mandate stricter standards in the Northeast and as a result of subsequent litigation challenging their authority to do so. In February 1994, the Ozone Transport Commission (OTC), which represents 12 Northeastern states and the District of Columbia, recommended that EPA require the adoption of California's LEV program in the Northeast to help improve air quality.[7] Subsequently, EPA issued a rule in January 1995 to implement the OTC's recommendation.[8] New York, Massachusetts, Vermont, and Maine have adopted California's LEV program. However, Virginia and several vehicle manufacturers legally challenged EPA's authority to mandate stricter standards. A U.S. Court of Appeals ruled that EPA does not have the authority under the Clean Air Act to impose stricter standards for new vehicles prior to MY2004 even if air quality problems warranted such action.[9] During these events, EPA held ongoing discussions with states and manufacturers to develop an alternative to mandating tighter standards, which resulted in an agreement to establish the National LEV program based on California's low emission requirements.

In addition to challenging EPA's authority to mandate stricter standards, manufacturers also were concerned about the possibility of many states adopting California's sales target for

levels and for maintairling air quality in areas that once were in nonattainment but have since met the standards For flirther discussion, refer to CRS Issue Brief 97007, *Clean Air Act Issues,* by James McCarthy.

[5] American Honda Motor Company, American Suzuki Motor Corporation, BMW of North America, Chrysler Corporation, Fiat Auto U.S.A., Ford Motor Company, General Motors Corporation, Hyundai Motor America, Isuzu Motors America, Jaguar Motors Ltd., Kia Motors America, Land Rover North America, Mazda ~orth America), Mercedes-Benz of North America, Mitsubishi Motor Sales of America, Nissan North America, Porsche Cars of North America, Rolls-Royce Motor Cars, Saab Cars USA, Subarn of America, Toyota Motor Sales, U.S.A., Volkswagen of America, and Volvo North America Corporation.

[6] The eight Northeastern states include: Connecticut, Delaware, Maryland, New Hampshire, New Jersey, Pennsylvania, Rhode Island, and Virginia.

[7] Section 184 of the Clean Air Act established an interstate ozone transport region in the Northeast, which includes Connecticut, Delaware, Maine, Maryland, Massachusetts, New Hampshire, New Jersey, New York, Pennsylvania, Rhode Island, Vermont, Virginia, and the District of Columbia. The Act also directed EPA to establish a commission composed of representatives from these states. The commission's purpose is to determine the degree to which the transport of ozone and its precursors occurs throughout the region, assess strategies for mitigating interstate ozone transport, and recommend measures to assist states in the region with attaining the federal air quality standard for ozone.

[8] EPA. *Federal Register.* January 24, 1995. p.4712-4739.

[9] D.C. Circuit Court of Appeals. VIrginia vs. EPA. No 95-1163. March 11, 1997.

zero emission vehicles (ZEVs). Industry generally views ZEV sales targets as unrealistic because the costs of electric and fuel cell technologies necessary to eliminate all emissions remain significantly higher than conventional gasoline and diesel engines and the market for vehicles using these technologies is not yet established. As a means to encourage the participation of manufacturers, the National LEV program does not include ZEV sales targets.

This report provides background information on federal emission standards for motor vehicles under the Clean Air Act and stricter standards originally developed to address the severity of air quality problems in California, explains low emission standards and flexible compliance mechanisms to which states and manufacturers have voluntarily agreed under the National LEV Program, discusses EPA's estimates of air quality benefits and costs, and examines regulatory issues related to its implementation including sulfur levels in gasoline and the relative stringency of emission standards for light trucks.

Current Regulation of Vehicle Emissions

The Clean Air Act Amendments of 1990 (P.L. 101-549) established more stringent federal emission standards for all new vehicles manufactured for sale in the United States. The current standards, referred to as Tier 1, replaced less stringent requirements previously introduced under the Clean Air Act Amendments of 1977 (P.L. 95-95). While the Clean Air Act generally prohibits individual states from enforcing vehicle emission standards that are different from the federal ones, it provides an exception that allows California to develop stricter requirements to help address serious air quality problems in many of that state's metropolitan areas. The Tier 1 standards and California's more stringent requirements are discussed below.

Tier 1 Standards

The Tier 1 standards regulate emissions of non-methane hydrocarbons (NMHC), CO, NOx, and PM from each new vehicle manufactured for sale in the United States.[10] To reduce the regulatory burden on manufacturers, the standards were phased in gradually over three years beginning in MY1994. Passenger automobiles and smaller light trucks weighing up to 3,750 pounds are subject to more stringent standards than larger light trucks (including many sport-utility vehicles and minivans) weighing between 3,751 and 5,750 pounds. Heavier duty light trucks weighing more than 5,750 pounds are subject to the least stringent standards. In addition, each weight category is subject to separate standards for different intervals of vehicle use: an intermediate operational life of 5 years or 50,000 miles and a full operational life of 10 years or 100,000 miles (11 years or 120,000 miles for light tucks weighing more than 5,750 pounds). Section 202 of the Clean Air Act requires EPA to determine whether air quality needs warrant revisions to the Tier 1 standards to achieve further reductions in emissions. However, more stringent Tier 2 standards could not become effective prior to MY2004. EPA submitted its Tier 2 study to Congress in August 1998, and if it deems

[10] Clean Air Act, Section 202. 42 U.S.C. 7521.

necessary, is required by the Clean Air Act to propose such standards by December 31, 1999 through a rulemaking process.[11]

California's Low Emission Standards

EPA approved California's LEV program in January 1993 with authority provided under Section 209 of the Clean Air Act.[12] The program became effective in MY1994 and requires manufacturers to comply with low emission standards for vehicles produced for sale in the state. It retains the basic structure of the Tier 1 requirements, under which passenger automobiles and light trucks are subject to separate sets of standards according to vehicle weight and amount of use. In addition to stricter standards, California's program differs from the Tier 1 requirements in three ways. First, it replaced the testing procedure for NMHC with a test for non-methane organic gases (NMOG) that more accurately measures the level of hydrocarbons, and it added a new standard for formaldehyde (HCHO), a carcinogen. Second, manufacturers must certify each vehicle under one of five categories: California Tier 1 vehicles, transitional low emission vehicles (TLEVs), low emission vehicles (LEVs), ultra low emission vehicles (ULEVs), or ZEVs. However, they have the flexibility to demonstrate compliance by using fleet emission averages instead of certifying each vehicle according to the same stringency. Third, the program requires that ZEVs comprise 10% of total vehicle sales in the state by MY2003. On November 5, 1998, the California Air Resources Board, a state agency, finalized revisions to its LEV program (LEV II) that will become effective in MY2004 and increase the stringency of that state's low emission standards beyond their current level.[13] California's current LEV I standards will remain in effect through MY2003.

National Low Emission Vehicle Program

The National LEV program adopts California's current LEV I standards for NMOG, NOx, CO, PM, and HCHO.[14] It also incorporates flexible compliance mechanisms that allow manufacturers to use fleet emission averages instead of certifying each vehicle according to the same stringency. The program focuses on controlling the formation of ozone by requiring a 70% reduction NMOG emissions and a *50%* reduction in NO_x emissions. The standards require much fewer reductions in CO emissions primarily because ambient concentrations of this pollutant have been declining steadily since 1987. Gasoline-powered vehicles are not significant sources of PM, and the low emission standard for this pollutant is limited to diesel vehicles. The standard for HCHO would provide additional air quality benefits by reducing carcinogen levels. However, certified vehicles may emit greater quantities of pollutants in actual use than the program's standards indicate if the regulatory issues of gasoline sulfur

[11] EPA. Office of Air and Radiation. *Tier 2 Report to Congress.* July 1998. 55 p.

[12] EPA. *Federal Register.* January 13, 1993. p.4166. Section 209 of the Clean Air Act allows EPA *to* grant a waiver to California for adopting vehicle emission standards that are more stringent *than* the federal ones. Section 177 permits other states with serious air quality problems to adopt stricter standards if they are identical to California's.

[13] For information on California's new LEV II standards, refer to the California Air Resources Board Home Page at *http://www.arb.ca.gov*

[14] EPA. *Federal Register.* June 6, 1997. p.31192-31270.

levels and the relative stringency of emission standards for light trucks are not addressed, both of which are discussed later in this report. Specific requirements of the program and EPA's estimates of air quality benefits and costs are explained below.

Certification Standards

Similar to California's LEV program, manufacturers must certify each vehicle according to one of four different categories subject to increasingly stringent emission standards: federal Tier 1, TLEVs, LEVs, or ULBVs. While the National LEV program does not include California's ZEV sales target, manufacturers also can certif~ their vehicles as ZEVs and receive credit for their air quality benefits. Relative to the NMOG standard, vehicles certified as TLEVs will be 50% cleaner than Tier 1 vehicles, those certified as LEVs will be 70% cleaner, and those certified as ULEVs will be 85% cleaner. Like the Tier 1 and California standards, passenger automobiles and smaller light truclcs weighing up to 3,750 pounds are subject to stricter standards than larger light trucks weighing between 3,751 and 5,750 pounds. (*Refer to table.*) Light trucks weighing more than 5,750 pounds remain subject to the less stringent Tier 1 standards, as there are no low emission standards for these heavier vehicles. In addition, certified vehicles also must other low emission standards for the Supplemental Federal Test Procedure which measures emissions during high speed acceleration, high speed driving, and air conditioning use.

Fleet Average Standards

While certified vehicles are subject to emission standards for a total of five pollutants, manufacturers will demonstrate compliance by averaging only the NMOG emissions of their vehicle fleets, which represent the most significant reductions under the program. Similar to the certification standards, passenger automobiles and smaller light trucks weighing up to 3,750 pounds are subject to stricter fleet averages than larger light trucks weighing between 3,751 and 5,750 pounds. Vehicles sold in the Northeast in MY1999 and MY2OOO will be subject to less stringent averages than those that will become effective nationwide in MY2OO1. At this time, the fleet average standards will become as stringent as the LEV certification standards for NMOG at 5 years or 50,000 miles. (*Refer to table.*) To preserve regional air quality benefits, manufacturers must calculate separate fleet emission averages for vehicles sold in the Northeast and for those sold in other states.

Trading Program

To provide additional flexibility in demonstrating compliance, manufacturers can trade NMOG credits to meet the fleet average standards. A manufacturer will generate credits in a model year if its fleet average is lower than the standard, and will generate debits if its fleet average exceeds the standard. Manufacturers with debits can purchase credits from others to achieve compliance. However, trading will be conducted in two separate regions (the Northeast and the remaining states) to prevent manufacturers from using credits generated in states with relatively clean air to achieve compliance in the more heavily polluted

Northeastern states. Manufacturers also can save credits for use or sale in future model years. However, these credits will lose a portion of their value each year and will no longer be valid after the third year held. As an incentive to introduce cleaner vehicles nationwide sooner than the program requires, manufacturers will be eligible for early reduction credits if they meet the fleet average standards in states outside of the Northeast prior to MY2001.

EPA's Estimates of Air Quality Benefits and Costs

The main air quality benefit of the National LEV program would be reductions in the ozone precursors, NMOG and NOx, which also could help to reduce fine particulates. Other benefits would be reductions in two carcinogens, formaldehyde and benzene.[15] Based on the stringency of the certification standards, EPA estimates that by 2005 the National LEV program would reduce NO_x emissions by 400 tons per day nationwide, NMOG by 279 tons, PM by 29 tons, benzene by 7 tons, and formaldehyde by 4 tons.[16] While such reductions would help to reduce pollution from new vehicles, they are relatively small (less than 1%) compared to total emissions of these pollutants from all sources combined. In 1996, nationwide emissions of NO_x were roughly 64,000 tons per day, emissions of VOCs (including NMOG) were over 52,000 tons per day, and emissions of PM were approximately 9,000 tons per day.[17] Actual emission reductions will depend to a large extent on whether certified vehicles operate as cleanly in actual use as they do under test conditions.

EPA analyzed the costs of the National LEV program based on cost estimates of vehicles sold in California. The California Air Resources Board estimates that the cost of an LEV in the state is on average $96 higher than a Tier I vehicle. EPA estimates a somewhat lower cost of $76 under the National LEV program due to the following reasons: advances in technology could reduce costs; economies of scale also could lower costs since manufacturers will be able to produce a single fleet of vehicles subject to the same standards nationwide; past cost estimates from the California Air Resources Board have proven higher than actual price differences; and industry historically has lowered vehicle prices in successive model years after the introduction of new technologies. Based on projections of vehicle sales nationwide, EPA estimates that the total cost of the National LEV program would be $950 million annually.[18]

---

[15] Although the National LEV program does not include a standard for benzene, EPA expects that the low emission technologies used to achieve the certification standards also would reduce emissions of this pollutant.

[16] EPA. *Federal Register.* June 6, 1997. p.31195-31196.

[17] EPA. Office of Air Quality Planning and Standards. *1996 National Air Quality and Emissions Trends Report.* January 1998. p.84-85. EPA estimates that in 1996 stationary and mobile sources in the United States emitted 23.4 million tons of NO_x, 19.1 million tons of VOCs, and 3.3 million tons of PM, yielding adaily emissions average of roughly 64,000 tons for NO_x, over 52,000 tons for VOCs, and approximately 9,000 tons for PM.

[18] EPA. Federal Register. June 6, 1997. p. 31197

Comparison of Tier 1 and National LEV Standards

Vehicle Type / Weight	5 Years / 50,000 miles (grams/mile)					10 Years / 100,000 miles (grams/mile)				
	NMHC/ NMOG **a**	Co	NOx	PM **b**	HCHO	NMHC/ NMOG a	CO	NOx	PM **b**	HCHO
Tier 1 Certification Standards										
LDVs/LDTs (0-3750 lbs.)	.250	3.4	.4	.08	--	.310	4.2	.60	.10	--
LDTs (3751-5750 lbs.)	.320	4.4	.7	.08	---	.400	5.5	.97	.10	--
National LEV Certification Standards										
LDVs/LDTs (0-3750 lbs.)										
TLEV	.125	3.4	.4	--	.015	.156	4.2	.60	.08	.018
LEV	.075	3.4	.2	--	.015	.090	4.2	.30	.08	.018
ULEV	.050	1.7	.2	--	.008	.055	2.1	.30	.04	.011
LDTs (3751-5750 lbs.)										
TLEV	.160	4.4	.7	--	.018	.200	5.5	.90	.10	.023
LEV	.100	4.4	.4	--	.018	.130	5.5	.50	.10	.023
ULEV	.050	2.2	.4	--	.009	.070	2.8	.50	.05	.013

Vehicle Type / Weight	5 Years / 5O,000 miles (grams/mile)					10 Years / 100,000 miles (grams/mile)				
National LEV NMOG Fleet Averages										
LDVs/ILDTs (0-3750 lbs.)										
MY1999(Northeast)	.148	--	--	--	--	--	--	--	--	--
MY2000 (Northeast)	.095	--	--	--	--	--	--	--	--	--
MY2001 (Nationwide)	.075	--	--	--	--	--	--	--	--	--
LDTs (3751-5750 lbs.)										
MY1999 (Northeast)	.190	--	--	--	--	--	--	--	--	--
MY2OOO (Northeast)	.124	--	--	--	--	--	--	--	--	--
MY2001 (Nationwide)	.100	--	--	--	--	--	--	--	--	--

[a] Tier 1 vehicles are subject to the testing procedure for NMH C, whereas vehicles certified under the National LEV program are subject to the testing procedure for NMOG. Both pollutants are closely related hydrocarbon compounds that can combine with NO_x in the presence of sunlight to form ozone. The National LEV program adopted California's use of the testing procedure for NMOG because it more accurately measures the level of hydrocarbons in a vehicle's exhaust.

[b] The National LEV standards for PM apply only to diesel-fueled vehicles at 10 years or 100,000 miles.

c The National LEV program requires manufacturers to meet fleet average standards only for NMOG at 5 years or 50,000 miles.

Prepared by the Congressional Research Service with data from the Environmental Protection Agency.

LDV = light duty (passenger) vehicle
TLEV = Transitional low emission vehicle
ULEV = ultra low emission vehicle
NMOG = non-methane organic gases
NO_x = nitrogen oxides
HCHO = formaldehyde

LDT = light duty truck
LEV = low emission vehicle
NMHC = non-methane hydrocarbons
CO = carbon monoxide
PM = particulate matter

RELATED REGULATORY ISSUES

Two regulatory issues could play significant roles in determining the extent to which the National LEV program is able to reduce actual emissions and help states improve air quality. First, certified vehicles may not operate as cleanly in actual use as the program's low emission standards suggest because high sulfur levels in commercial gasoline could diminish the effectiveness of low emission technologies. Second, light trucks are not regulated as strictly as passenger automobiles. As a result, the increasing market share of light trucks could alter the composition of the vehicle fleet significantly enough to increase total emissions and offset the sharper reductions from passenger automobiles. Each of these issues is discussed below.

Sulfur Levels in Gasoline

Sulfur occurs naturally in crude oil, and the amount remaining in gasoline varies widely among refineries. Nationwide, the sulfur level in commercial gasoline averages 330 parts per million (ppm) and can reach as high as 1,000 ppm in some areas.[19] Levels are lowest in California where a cap is currently in place that limits sulfur to a maximum of 80 ppm with a statewide average of 30 ppm. To meet emission standards, conventional gasoline-powered vehicles primarily depend on catalytic converters to reduce levels of CO, NO_X, and VOCs from the exhaust. However, the sulfur present in gasoline can chemically react with the surface of the catalyst and inhibit the removal of other pollutants. The extent to which sulfur increases emissions depends on numerous factors such as the design of the catalyst, materials used, air to fuel mixture of the engine, and range of exhaust temperatures.

Currently, a substantial discrepancy exists between sulfur levels in commercial gasoline and the amount in fuels used to certiiy that vehicles meet emission standards. Under the National LEV program, manufacturers can certify their vehicles with the test fuel for California's LEV program or the federal test fuel for Tier 1 vehicles. Sulfur levels in California's test fuel are frequently below 30 ppm, and the amount present in federal test fuel is typically much less than 100 ppm, both of which are lower than the average of 330 ppm in commercial gasoline available nationwide. Even in areas where the second phase of reformulated gasoline will be sold in 2000, sulfur levels will average as much as 150 ppm and continue to exceed test fuel amounts by a significant margin.[20] In effect, vehicles certified under the National LEV program may emit greater quantities of pollutants in actual use than the stringency of the low emission standards alone would suggest, and therefore, make less of an impact on improving air quality. However, the extent to which the air quality benefits of the program might be compromised is difficult to determine since different makes and models of vehicles can respond quite differently to high sulfur gasoline.

The automobile and oil industries have conducted studies to examine the effects of gasoline sulfur levels on emissions performance. They generally conclude that actual emissions depend on the amount of sulfur in gasoline, and NO_x emissions seem to be affected

---

[19] EPA. Office of Mobile Sources. *StaffPaper on Gasoline Sulfur Issues*. May 1998, p.3.

[20] Reformulated gasoline (RFG) burns more cleanly and results in fewer emissions ofozoneforming and toxic compounds. Under the Clean Air Act, 10 metropolitan areas with serious air quality problems require the sale of RFG, and 13 additional areas voluntarily require the sale of RFG. The majority of these areas are located in Northeastern states.

by sulfur more dramatically than other pollutants. However, some vehicles are not as sensitive to high sulfur levels as others. EPA has combined the test results from two recent industry studies that indicate a wide range of effects among LEVs and ULEVs when operated on high sulfur gasoline. For example, emissions of NO_x from the 9 most sensitive vehicles increased by 251% when operated on gasoline with a national sulfur average of 330 ppm, whereas NO_x emissions from the 9 least sensitive vehicles increased by only 61% when operated on the same gasoline. These studies also tested the effects of sulfur on Tier I vehicles and observed a much lower increase in NO_x emissions of 13.6% relative to the national sulfur average of 330 ppm, which indicates that the technologies used to achieve the LEV and ULEV standards are significantly more sensitive than those typically used in Tier 1 vehicles.[21]

The effects of gasoline sulfur levels on emissions could possibly be addressed in two major ways. First, manufacturers might be able to redesign vehicle engines and exhaust systems so that they are less sensitive to sulfur. Second, the amount of sulfur in gasoline might be limited either nationally or on a regional basis in areas with serious air quality problems. While the technology is currently available to lower sulfur levels at least to that already achieved in California, such reductions would require refineries to invest in new equipment and lead to increases in operating costs of 5 to 8 cents per gallon, which in turn could raise gasoline prices.[22] However, recent developments in refinery technology could lower the costs of reducing gasoline sulfur levels to 1 or 2 cents per gallon nationwide. If EPA proposes more stringent Tier 2 standards, it also may consider regulatory controls on gasoline sulfur levels to insure that the low emission technologies necessary to achieve further reductions in emissions from new vehicles are not compromised.

Relative Stringency of Emission Standards for Light Trucks

The discrepancy between the standards for passenger automobiles and light trucks is a rising issue in the regulation of vehicle emissions. While smaller light trucks weighing up to 3,750 pounds are subject to the same emission standards as passenger automobiles, larger light trucks weighing in excess of this amount are allowed to meet less stringent standards. Congress initially established more lenient standards for larger light trucks because their capacity for carrying heavy loads compromised emissions performance and they composed a smaller portion of the vehicle fleet, which made less of an impact on air quality. However, the composition of the vehicle fleet has changed over the years in favor of light trucks, and their market share has increased from 16.5% in 1980 to 40.4% of total vehicle sales in 1996.[23] The sharp rise in sales is mostly due to the heightened popularity of sport-utility vehicles and minivans. Such vehicles accounted for only 13.6% of total light truck sales in 1975, but by

[21] EPA. Office of Mobile Sources. *Staff Paper on Gasoline Sulfur Issues.* May 1998. p.11. Test results compiled from two studies conducted by the Coordinating Research Council (CRC), representing the oil industrv, and by the American Automobile Manufacturers Association and the Association of International Automobile Manufacturers.

[22] Ibid., p.32-34. EPA and the Department of Energy have studied the costs of reducing sulfur levels in gasoline to an average of 40 ppm in the Northeast and Gulf Coast states and have estimated that costs to refmeries would range between 5.1 and 8 cents per gallon. The American Petroleum Inntitute sponsored a different study to also include the Midwest, which estimated an average cost to refineries of no more than 5.1 cents per gallon.

[23] Department of Transportation. Bureau of Transportation Statistics. *Transportation Statistics Annual Report for FYI 997* p.87

1996, their market share had more than tripled to nearly 50% of all light trucks sold nationwide.[24] Many of these vehicles weigh in excess of 3,750 pounds and are therefore allowed to meet less stringent emission standards even though they are used primarily for personal transportation rather than for carrying heavy loads, which the more lenient standards were initially designed to accommodate. As a result, the larger market share of light trucks and increased mileage of these vehicles from passenger use has expanded the total emissions inventory of the on-road vehicle fleet and raised the issue of whether light trucks should be regulated more strictly.

Like the Tier 1 standards under the Clean Air Act and California's current low emission standards, the National LEV program continues the practice of allowing light trucks weighing in excess of 3,750 pounds to meet less stringent standards. Consequenily, if the market share of higher-emitting light trucks continues to rise, the air quality benefits of the National LEV program might be less promising. If EPA proposes more stringent Tier 2 standards, it might regulate emissions from all light trucks more strictly due to air quality considerations. Higher emissions from light trucks could possibly be addressed in at least three different ways. First, all light trucks might become subject to the same standards as passenger automobiles regardless of vehicle use. Second, larger light trucks might continue to be subject to less stringent standards but be required to use the same low emission technologies as passenger automobiles. Third, separate standards might be established according to vehicle use rather than vehicle type. While the first option has the greatest potential for providing the most air quality benefits, reducing emissions from heavier trucks with larger engines could prove to be problematic as it would require manufacturers to use more advanced catalytic technologies that could increase production costs and to improve engine operating efficiencies which might result in diminished performance. As discussed earlier, California recently finalized revisions to its LEV program (LEV II) and decided to address the issue of higher emissions from light trucks by requiring all vehicles weighing up to 8,500 pounds to meet the same emission standards beginning in MY2004.

CONCLUSION

The automobile industry has been complying with low emission standards in California for the past few years. During this time, manufacturers have demonstrated that the technology to produce cleaner vehicles is currently available. However, the ability of this technology to reduce emissions in actual use depends on how sensitive a vehicle is to the amount of sulfur in gasoline. At this time, low sulfur gasoline is not commercially available outside of California, and plans under the National LEV program to reduce vehicle emissions with current technologies in areas where cleaner fuels are not available could prove extremely difficult. In addition, the rising market share of light trucks could result in expanding the emissions inventory of the nation's on-road vehicle fleet and offset some of the sharper reductions in emissions from passenger automobiles. While the National LEV program will regulate emissions from new vehicles more strictly than the current standards under the Clean Air Act, it is a partial artempt at introducing cleaner vehicles nationwide because it leaves the two larger regulatory issues of gasoline sulfur levels and higher emissions from light trucks

[24] Ibid., p. 88

open for further debate. EPA may address these issues in the near future if it decides to propose more stringent Tier 2 emission standards.

Chapter 10

AIR QUALITY AND EMISSIONS TRADING: AN OVERVIEW OF CURRENT ISSUES*

David M. Bearden

INTRODUCTION

Conventional air quality regulations place fixed limits on emissions from individual sources and require them to install specific technologies to control pollution. However, compliance may be more economically feasible for some sources than it is for others. Emissions trading is a market-based alternative to conventional regulation which permits sources facing high pollution control costs to meet their emission limits by purchasing excess reductions from other sources that can afford to lower their emissions further than federal or state regulations require. Trading has the potential to improve air quality in cases where a pollutant disperses over a broad geographic area and the environmental objective is to control total emissions rather than limit local emissions from individual sources. In the United States, experience with trading programs began in the 1970's when the Environmental Protection Agency (EPA) used its authorities under the Clean Air Act to develop more flexible policies on emission control measures that states could pursue to comply with federal air quality standards. Most trading programs have focused on pollutants generated from large stationary sources that account for a significant share of total emissions, such as electric utility plants and industrial boilers that burn fossil fuels.[1]

The following sections of this report explain how trading programs work, analyze factors that can influence their effectiveness, discuss current and potential uses of trading to improve air quality, and examine issues relevant to their implementation.

* Excerpted from CRS Report RS20891. Updated August 16, 1999.

[1] For a broader discussion, refer to CRS Report 94-213.

How Do Trading Programs Work?

Trading programs use either credits or allowances. Credits are emission reductions that a pollution source has achieved in excess of required amounts. Sources that have earned credits can sell them to others that need additional reductions. Allowances differ from credits in that they represent the amount of a pollutant that a source is permitted to emit during a specified time in the future. If a source estimates that its emissions will be less than its allowances, it can sell its excess allowances to other sources that need them. Some trading programs also allow sources to save credits or allowances for meeting limits on emissions in future years. This practice is commonly referred to as *banking*.

What Factors Can Influence the Effectiveness of Trading?

For diffuse pollutants, decreasing total emissions over large areas through trading has the potential to improve overall air quality. For example, trading is well suited for pollutants such as nitrogen oxides (NOx), sulfur oxides (SOx), and carbon dioxide (CO_2) because they can drift far from their points of release.[2] However, trading is less suited for pollutants that remain locally concentrated because excess reductions in one area would not improve air quality elsewhere, and localized problems with air quality could arise from trading if sources clustered in one area are permitted to offset substantial increases in pollution with credits or allowances. Whether trading can improve overall air quality depends on many factors related to program design, implementation, and enforcement.

Defining the Trading Area

The geographic area within which trading occurs depends on how far a pollutant travels from its points of release. For example, if a pollutant disperses over a region of states, sources in different states could trade emission reductions and improve overall air quality. However, if emissions remain largely within a state or a portion of a state, trading would need to be confined within these boundaries because reductions in other locations would not improve air quality in the affected area.

Capping Emissions

Placing a cap (i.e., a limit) on total emissions from all participating sources can help to protect overall air quality regardless of the volume of trading because the aggregate level of pollution would not exceed a predetermined amount. However, using a cap can be a complicated issue for at least two reasons. First, a limit on total emissions must be determined which would accomplish a particular environmental objective. Second, a formula must be developed for distributing the amount of the cap among the pollution sources involved.

---

[2] NO_x is a precursor to ozone, and along with SO_x, can contribute to the formation of acid rain. $C0_2$ is a greenhouse gas that may contribute to global warming.

Determining Compliance

A system to record each trade and monitor emissions is necessary to determine whether the pollution that a source generates exceeds the credits or allowances that it holds. However, if the transaction process to record trades is too complex or the cost of monitoring equipment is too high, determining compliance would be extremely difficult. Under these circumstances, trading might prove infeasible.

Penalizing Sources for Noncompliance

To provide an economic incentive for sources to comply with their emission limits, penalties for generating excessive pollution must he significantly higher than the cost of reducing emissions or purchasing reductions from another source. If penalties for exceeding emission limits are too low, sources facing high expenses to control pollution might prefer to pay a penalty that is less than the cost of purchasing credits or allowances to meet their limits.

Banking Credits or Allowances

Allowing pollution sources to *bank* (i.e., save) allowances or credits for use or sale in the future can provide an incentive to keep emissions below allowable levels in the short-term. Air quality may improve more quickly if enough sources bank rather than use their credits or allowances. However, such gains could diminish at a later point if sources use their banked reductions to increase emissions.

What Are Some Examples of Trading Programs?

While early trading programs were based mostly on credits, allowance programs with emissions caps have become more common. Although exact savings are difficult to estimate, recent experiences with allowance trading under the federal acid rain program and in California's efforts to control ozone in the Los Angeles area indicate that these programs likely have reduced overall emissions at lower costs than conventional regulation. Whether an allowance program to control ozone transport in 12 northeastern and midatlantic states, which began in May 1999, will achieve similar results depends in part on the cooperation of the states involved.[3] Each program is discussed below.

Federal Acid Rain Program

To combat acid rain, the Clean Air Act Amendinents of 1990 established an allowance program to reduce emissions of sulfur dioxide (S02). The program capped national emissions at roughly 50% of 1980 levels. Phase I of the program began in 1995 and involves over 400

[3] Ozone transport can result when precursors such as NO_x, drift across state borders and cause ozone levels to increase in areas where actual emissions are relatively low.

coal-fired electric utility generating units. Phase 11 begins in 2000 when the number of participating units will increase to over 2,000. EPA reports that actual emissions have been below the established cap each year from 199510 1998.[4] However, emissions have remained below the cap because utilities banked about 31% of total allowances during this time. Consequently, emissions could rise in coming years if utilities decide to use their banked allowances to achieve compliance. As the program enters its second phase, the substantial increase in the number of participating sources also could make the program more difficult to administer and perhaps reduce its effectiveness.

Regional Clean Air Incentives Market

In January 1994, California's South Coast Air Quality Management District introduced the Regional Clean Air Incentives Market (RECLAIM) in the Los Angeles area. RECLAIM is an allowance program aimed at reducing emissions of NO_x and SO_X The program's goal is to achieve an 80% reduction in emissions by 2003. Stationary sources that annually emit at least 4 tons of NO_x or SO_X receive allowances, which are based on past peak production and existing emission requirements. The total number of allowances is equal to the cap on emissions. The state reports that actual emissions have been below the cap each year from 1994 to l997.[5] However, some critics argue that actual reductions in emissions have been less dramatic than a comparison to the cap suggests because emission allowances were initially high.

Interstate Trading

In 1994, the Ozone Transport Commission, representing 12 midatlantic and northeastern states, the District of Columbia, and EPA, agreed to implement an interstate trading program that caps NO_x emissions at less than half of 1990 levels to assist the region in complying with the federal air quality standard for ozone.[6] The first phase of the program began in May 1999 and allows large electric utilities and industrial boilers within the region to trade NO_x allowances to meet their emission limits.[7] The second phase of the program is scheduled to begin in 2003 when the emissions cap will be lowered. EPA will record each trade and track monitoring data on actual emissions. Each participating state will be responsible for allocating allowances to pollution sources and enforcing program requirements within its borders. Since enforcement will be a state role, the effectiveness of the program largely will depend on the combined efforts of the states in carrying out their implementation responsibilities.

---

[4] EPA. Office of Air and Radiation. *1998 Acid Rain Program Compliance Report.* August 1999. The report is available electronically from EPA's web site at *http://www.epa.gov/acidrain* .

[5] South Coast Air Quality Management District. *RECLAIM Audit Report for the 1997 Compliance Year.* March 12, 1999. p 25.

[6] 12 states include Connecticut. Delaware, Maine, Maryland, Massachusetts, New Hampshire, New Jersey, New York, Pennsylvania, Rhode Island, Vermont, and Virginia. However, the state of Virginia chose not to participate in the interstate trading program.

[7] Addifional information on the Ozone Transport Commission's program is available electronically from EPA's web site at *http://www.epa.gov/acidrain/otc/otcmain.html* .

WHAT ARE SOME OF THE POTENTIAL USES OF TRADING?

While conventional regulation remains the most common measure to control air pollution, the early success of the federal acid rain program and RECLAIM in Califomia has encouraged the consideration of trading to address other air quality concerns. Domestically, trading may be used to reduce ozone transport across 22 eastern states, provide the private sector with flexibility in complying with EPA's Tier 2 proposal to reduce emissions from new motor vehicles, and control possible increases in pollution if the electric utility industry is restructured. Internationally, trading may be used to assist nations in reducing greenhouse gas emissions thought to contribute to global warming. Each of these issues is discussed below.

Regional Limits on Nitrogen Oxides

In October 1998, EPA finalized a rule that requires 22 states and the District of Columbia to submit state implementation plans by September 30, 1999 to address the regional transport of ozone.[8] The rule limits NO_x emissions from large electric utilities and other combustion sources. To provide flexibility in meeting emission limits, states have the option of participating in a regional cap and trade program.[9] In May 1999, a D.C. Circuit Court of Appeals decision delayed the implementation of the ozone transport rule while the court considers the merits of a legal challenge from midwestern states that EPA's limits on NO_x emissions are unnecessarily strict.[10] In light of the court's decision, negotiations between EPA and the affected states are underway to develop alternative emission controls that are acceptable to all parties.

Restructuring of the Electric Utility Industry

Several bills have been introduced in the 106th Congress that would restructure the electric utility industry with some of the possible outcomes being increased competition among generators and lower costs for consumers.[11] The possibility of increased competition has raised the issue of whether generators would rely more heavily on older utility plants to lower their costs, which typically generate significant emissions of NOx, SOX, and C02. Consequently, increases in pollution could complicate state efforts in complying with federal air quality standards. Seven bills (H.R. 25, H.R. 657, H.R. 1828, S. 172, S. 673, S. 1047, and S. 1369) have been introduced in the 106th Congress that would permit the use of trading to control air pollution from electric utilities.

---

[8] EPA. *Federal Register.* October 27,1998. p.57356-57538. The 22 states include: Alabama, Connecticut, Delaware, Georgia, Illinois, Indiana, Kentucky, Massachusetts, Maryland, Michigan, Missouri, North Carolina, New Jersey, New York, Ohio, Pennsylvania, Rhode Island, South Carolina, Tennessee, Virginia, Wisconsin, and West Virginia.

[9] For further discussion, refer to CRS Report 98-236.

[10] D.C. Circuit Court of Appeals, *Michigan vs. EPA,* May 25, 1999, No.98-1497.

[11] For a discussion of each bill, refer to CR5 Issue Brief IB10006.

EPA's Tier 2 Proposal

On May 13, 1999, EPA proposed more stringent Tier 2 emission standards for new motor vehicles.[12] The standards would be phased in between 2004 and 2009 and require vehicle manufacturers to reduce NO_x emissions by roughly 90% below levels currently required under the Clean Air Act. Manufacturers with NO_x fleet emission averages below .07 grams/mile would receive credits that could be used in later years or sold to other manufacturers having difficulty in meeting the standards.[13] Under EPA's proposal, oil refineries also would be required to reduce gasoline sulfur levels to an average of 30 parts per million beginning in 2004.[14] Refineries producing average sulfur levels below this amount would receive credits that could be banked for future use or sold to other refineries facing high costs to achieve compliance.

Economic Incentives for Telecommuting

Representative Wolf has introduced legislation (H.R. 2556) that would encourage the use of emissions trading to promote telecommuting as a measure to reduce traffic congestion and control vehicular pollution. Specifically, the bill would authorize $250,000 in FY2000 for awarding a grant to the National Environmental Policy Institute (a nonprofit organization) to design a pilot trading program in conjunction with the Departrnent of Transportation, Department of Energy, and EPA. The objective of the pilot program would be to test the feasibility of awarding tradeable credits for emission reductions attributed to telecommuting. The amount of the credits would be based on the extent to which the decline in the number of vehicle miles traveled reduces emissions. Regulated entities would be able to purchase such credits to assist them in complying with Clean Air Act requirements. The pilot program would be tested in five metropolitan areas with high levels of traffic congestion and poor air quality, including Los Angeles, California, and the District of Columbia.

Greenhouse Gas Reductions

The United States signed the Kyoto Protocol to the United Nations Framework Convention on Climate Change (UNFCCC) on November 12, 1998.[15] If approved for ratification by the Senate, it would commit the United States to reducing greenhouse gas emissions by 7% below 1990 levels. Although nations would be allowed to trade emission reductions to help meet their commitments, numerous issues remain unresolved, including the development of standards to ensure accurate monitoring and effective enforcement and how to address inequalities among nations in the resources that would be necessary to implement trading effectively.[16] The UNFCCC parties have set a deadline for resolving such issues by late 2000 or early 2001. Nations also might pursue domestic trading to assist them in meeting future commitments to reduce greenhouse gas emissions. As introduced, S. 547 and H.R.

---

[12] EPA. *Federal Register*. May 13,1999. p.26004-26142.

[13] For additional information on the proposed standards, refer to CRS Report RS20247.

[14] For further discussion, refer to CRS Report RS20163.

[15] For a summary of the agreement, refer to CRS Report 98-2.

[16] For further discussion, refer to CRS Issue Brief 97057.

2520 would authorize the President to enter agreements with industry and businesses to award tradeable credits for voluntarily reducing greenhouse gas emissions. Such credits would provide the private sector with greater flexibility in complying with regulatory limits that may be placed on greenhouse gas emissions in the future. However, some opponents argue that awarding early reduction credits would result in the premature implementation of the Kyoto Protocol. As introduced, H.R. 2221 would prohibit the use of federal funding to implement an early reduction credit program prior to the Senate's ratification of the Kyoto Protocol.

CONCLUSION

Numerous air quality problems have raised the issue of how to reduce emissions of certain pollutants while minimizing pollution control costs. Trading can be an attractive option due to the potential it offers for reducing emissions with greater flexibility and at less cost than conventional regulation. However, its effectiveness hinges upon the nature of the pollutant involved, program design, implementation, enforcement, and accurate monitoring data to verify that transactions between pollution sources represent actual reductions in emissions. The early success of the federal acid rain program and RECLAIM in California has encouraged the consideration of trading to address other air quality problems. Domestic trading to control ozone transport faces particular challenges because of the need for voluntary' cooperation by affected states to implement such programs effectively. International trading to reduce greenhouse gas emissions would face even more obstacles due in part to the need for uniform enforcement of program requirements and inequalities among nations in the resources essential for implementation.

Chapter 11

AIR QUALITY AND MOTOR VEHICLES: AN ANALYSIS OF CURRENT AND PROPOSED EMISSION STANDARDS*

David M. Bearden

INTRODUCTION

The Environmental Protection Agency (EPA) has proposed stricter controls on motor vehicle emissions and is currently implementing a voluntary program to introduce cleaner vehicles nationwide by model year (MY) 2001. Congress first required vehicle emissions to be regulated when it amended the Clean Air Act under the Motor Vehicle Air Pollution Control Act of 1965 (P.L. 89-272). While Congress subsequently amended the Clean Air Act in 1967, 1970, 1977, and 1990 to tighten controls on vehicle emissions, the Department of Transportation reports that the number of miles traveled by passenger cars and light trucks has risen from 1.0 trillion in 1970 to 2.3 trillion in 1996.[1] The rise in travel has offset much of the pollution reductions achieved by emission controls, and as a result, motor vehicles continue to be major sources of air pollution. EPA reports that in 1997 on-road vehicles accounted for about 58% of carbon monoxide (CO) emissions in the United States, nearly 30% of nitrogen oxides (NO_x), roughly 27% of volatile organic compounds (VOCs), and about 9% of particulate matter (PM).[2] NO_x and VOCs contribute to ground-level ozone pollution, commonly known as smog.[3]

Despite current pollution controls, areas in many states have not attained the National Ambient Air Quality Standards (NAAQS).[4] Attainment of the ozone standard has been a

---

* Excerpted from CRS Report RL30298. September 2, 1999.

[1] Department of Transportation. Bureau of Transportation Statistics. *National Transportation Statistics 1998.* [*http://www.bts.gov/btsprod/nts*]

[2] EPA. Office of Air Quality Planning and Standards. *1997 National Air Quality and Emissions Trends Report.* December 1998. p. 112-116.

[3] Vehicles do not directly emit ozone. Rather, NO_x and VOCs produce ozone when they chemically react in the presence of sunlight. Vehicles emit VOCs primarily in the form of hydrocarbons (HC).

[4] The NAAQS set safe ambient levels for carbon monoxide, lead, nitrogen dioxide, ozone, particulate matter, and sulfur dioxide. To date, EPA reports that 130 areas with a combined population of 113 million are in violation of the NAAQS for one or more pollutants. Nearly 100 million people, roughly 88% of the affected population,

persistent problem in the Northeast due to high levels of NO_x and VOCs. The Clean Air Act Amendments of 1990 (P.L. 101-549) established "Tier I" standards that placed stricter limits on tailpipe emissions from new motor vehicles beginning in MY1994. The law also required EPA to determine by December 31, 1999 if more stringent "Tier 2" standards would be necessary in MY2004 to attain or maintain the NAAQS, and if so, to promulgate them. However, before promulgation, EPA was to assess the availability and cost-effectiveness of pollution control technologies that would be necessary to achieve further reductions in emissions. In a report submitted to Congress in August 1998, EPA concluded that tougher standards are necessary and essential technologies are available and cost-effective.[5] As a result, EPA proposed stricter Tier 2 standards on May 13, 1999.[6] Subsequently, the agency held a series of public hearings from June 9-17, 1999 in Philadelphia, Atlanta, Denver, and Cleveland, and received additional public comment until August 2, 1999. EPA is planning to issue a final rule by the end of the year.

As proposed, the Tier 2 standards would be phased in between MY2004 and MY2009. All new passenger cars and light trucks up to 6,000 pounds would become subject to the standards by MY2007, but heavier trucks between 6,001 and 8,500 pounds (including some sport utility vehicles and minivans) would not be required to meet them until MY2009. Because sulfur in gasoline impairs the operation of emission control technologies, the proposal also would limit gasoline sulfur levels to an average of 30 parts per million (ppm) nationwide beginning in 2004, roughly 90% below the current national average of 340 ppm.[7] In congressional testimony and in comments submitted to EPA, the oil refining industry claims that such limits on sulfur levels would be too costly and unattainable by 2004, whereas the automobile industry insists that low sulfur gasoline is necessary to meet the proposed emission standards for motor vehicles. EPA may address these and other issues in its final rule.

Prior to EPA's Tier 2 proposal, 8 northeastern states and 23 vehicle manufacturers completed a voluntary agreement in March 1998 to develop a National Low Emission Vehicle (LEV) Program aimed at achieving additional emission reductions before MY2004.[8] Under the National LEV program, manufacturers began to produce cleaner vehicles for sale in the Northeast and the District of Columbia in MY1999 and will expand the program nationwide in MY2001.[9] The program uses the same low emission standards that are currently in effect in California, New York, Massachusetts, Vermont, and Maine. The main incentive for manufacturers to participate in the National LEV program is that they are able to

---

reside in ozone nonattainment areas. For a discussion of air quality attainment issues, refer to CRS Issue Brief IB 10004, *Clean Air Act Issues in the 106th Congress,* by James McCarthy

[5] EPA. Office of Air and Radiation. *Tier 2 Report to Congress.* July 31, 1998. 55 p.

[6] EPA. *Federal Register.* May 13, 1999. p. 26004-26142. The proposed regulations are available electronically from EPA at [*http://www.epa.gov/oms/tr2home.htm*].

[7] For further discussion of EPA's gasoline sulfur proposal, refer to CRS Report RS20163, *Sulfur in Gasoline,* by Stephen Thompson and James McCarthy.

[8] EPA. *Federal Register.* June 6, 1997. p. 31192-31270. The 23 vehicle manufacturers include American Honda Motor Company, American Suzuki Motor Corporation, BMW of North America, Chrysler Corporation, Fiat Auto U.S.A., Ford Motor Company, General Motors Corporation, Hyundai Motor America, Isuzu Motors America, Jaguar Motors Ltd., Kia Motors America, Land Rover North America, Mazda (North America), Mercedes-Benz of North America, Mitsubishi Motor Sales of America, Nissan North America, Porsche Cars of North America, Rolls-Royce Motor Cars, Saab Cars USA, Subaru of America, Toyota Motor Sales, U.S.A., Volkswagen of America, and Volvo North America Corporation.

[9] The eight Northeastern states are Connecticut, Delaware, Maryland, New Hampshire, New Jersey, Pennsylvania, Rhode Island, and Virginia.

produce a single fleet of vehicles for sale nationwide rather than having to produce a separate fleet to meet stricter requirements in only a few states. While the National LEV standards are more stringent than the current Tier I standards, they demand fewer reductions than EPA's Tier 2 proposal would require. For example, the National LEV standards require a 50% reduction in overall NO_x emissions, whereas the proposed Tier 2 standards would require overall emissions of this pollutant to be reduced by 88% to 95%.

While the National LEV program is based on a voluntary agreement, it arose out of attempts by a few states and EPA to mandate stricter standards in the Northeast prior to MY2004 and as a result of subsequent litigation challenging their authority to do so. In February 1994, the Ozone Transport Commission (OTC), which represents 12 northeastern states and the District of Columbia, recommended that EPA require the adoption of California's LEV program in the Northeast to help improve air quality.[10] Subsequently, EPA issued a rule in January 1995 to implement the OTC's recommendation.[11] New York, Massachusetts, Vermont, and Maine have adopted California's LEV program. However, Virginia and several vehicle manufacturers legally challenged EPA's authority to mandate stricter standards. A federal court ruled that EPA does not have the authority under the Clean Air Act to impose stricter standards for new vehicles prior to MY2004 even if air quality problems warrant it.[12] As an alternative, EPA negotiated a voluntary agreement with states and manufacturers to establish the National LEV program based on California's current low emission standards.

In addition to challenging EPA's authority to mandate stricter standards, manufacturers also were concerned that many states with poor air quality would adopt California's sales target for zero emission vehicles (ZEVs). To encourage the participation of manufacturers, the National LEV program does not include ZEV sales targets. Industry views such targets as unrealistic because the costs of electric and fuel cell technologies necessary to eliminate all emissions remain significantly higher than conventional gasoline and diesel engines and the market for vehicles using these technologies is not yet established. Due to similar concerns, EPA's Tier 2 proposal would not impose a ZEV sales mandate but would offer credits for producing such vehicles, which could be used to help meet compliance requirements.

Implementation of the National LEV program is currently underway, and oversight is a possibility in the 106th Congress as the program expands beyond the Northeast in MY2001. Congressional interest in EPA's Tier 2 proposal has focused primarily on the potential regulation of gasoline sulfur levels, and there has been both opposition and support among Members of Congress. The following sections of this report provide background information on the regulation of vehicle emissions in the United States, analyze key elements of the National LEV program and EPA's Tier 2 proposal, summarize major views on the proposal that have been expressed by the automobile industry, the oil refining industry, and some environmental organizations, and discuss relevant legislative activity in the 106th Congress.

[10] Section 184 of the Clean Air Act established an interstate ozone transport region in the Northeast, which includes Connecticut, Delaware, Maine, Maryland, Massachusetts, New Hampshire, New Jersey, New York, Pennsylvania, Rhode Island, Vermont, Virginia, and the District of Columbia. The Act also directed EPA to establish a commission composed of representatives from these states. The commission's purpose is to determine the degree to which the transport of ozone and its precursors occurs throughout the region, assess strategies for mitigating interstate ozone transport, and recommend measures to assist states in the region with attaining the federal air quality standard for ozone.

[11] EPA. *Federal Register*. January 24, 1995. p. 4712-4739.

[12] D.C. Circuit Court of Appeals. *Virginia vs. EPA*. No. 95-1163. March 11, 1997.

Current Regulation of Vehicle Emissions

The Clean Air Act Amendments of 1990 (P.L. 101-549) established "Tier I" standards that placed stricter limits on tailpipe emissions from new motor vehicles. The Tier I standards replaced less stringent requirements previously introduced under the Clean Air Act Amendments of 1977 (P.L. 95-95). While the Clean Air Act prohibits individual states from enforcing vehicle emission standards of their own, it allows California to enforce stricter requirements to help address serious air quality problems in many of that state's metropolitan areas. In addition, other states with poor air quality can choose to adopt California's standards but cannot develop different standards of their own. To date, New York, Massachusetts, Vermont, and Maine have adopted California's standards. The federal Tier I standards and California's more stringent requirements are discussed below.

Federal Tier I Standards

The Tier I standards regulate emissions of non-methane hydrocarbons (NMHC), CO, $N0_x$, and PM from each new vehicle manufactured or imported for sale in the United States.[13] To reduce the regulatory burden on manufacturers, the standards were phased in gradually over 3 years beginning in MY 1994. Light trucks between 3,751 and 5,750 pounds are subject to standards that are roughly 1.5 times less stringent than those for cars and light trucks up to 3,750 pounds. Heavier trucks weighing more than 5,750 pounds are subject to the most lenient standards, which are about 2.5 times less stringent than those for lighter vehicles up to 3,750 pounds. In addition, each weight category is subject to separate standards for different intervals of vehicle use: an intermediate useful life of 5 years or 50,000 miles and a full useful life of 10 years or 100,000 miles (II years or 120,000 miles for heavier tucks weighing more than 5,750 pounds). (Refer to Table I on page 7.)

California's Low Emission Standards

EPA approved California's LEV I program in January 1993 with authority provided under Section 209 of the Clean Air Act.[14] The program became effective in MY 1994 and requires the automobile industry to comply with low emission standards for vehicles produced for sale in the state. It retains the basic structure of the Tier I requirements, under which passenger cars and light trucks are subject to separate sets of standards according to vehicle weight and amount of use. In addition to stricter standards, California's program differs from the Tier I requirements in three ways. First, it replaced the NMHC testing procedure with a test for non-methane organic gases (NMOG) that more accurately measures hydrocarbons, and it added a new standard for formaldehyde (HCHO), a carcinogen. Second, manufacturers must certify each vehicle under one of five categories: California Tier I

---

[13] Clean Air Act, Section 202. 42 U.S.C. 7521.

[14] EPA. *Federal Register.* January 13, 1993. p. 4166. Section 209 of the Clean Air Act allows EPA to grant a waiver to California for adopting vehicle emission standards that are more stringent than the federal ones. Section 177 permits other states with serious air quality problems to adopt stricter standards if they are identical to California's.

vehicles, transitional low emission vehicles (TLEVs), low emission vehicles (LEVs), ultra low emission vehicles (ULEVs), or zero emission vehicles (ZEVs). However, manufacturers have the flexibility to demonstrate compliance by using fleet emission averages instead of certifying each vehicle according to the same stringency. The use of fleet emission averages allows manufacturers broad flexibility in selecting which set of standards apply to different models of vehicles, while still meeting the air quality objective of lowering overall emissions to a specific level. Third, the program requires that ZEVs comprise 10% of total vehicle sales in the state by MY2003. On November 5, 1998, the California Air Resources Board, a state agency, finalized revisions to its current LEV I standards, which will become effective in MY2004.[15] The LEV II standards will require further reductions in emissions and subject all new passenger cars and light trucks up to 8,500 pounds to the same regulations. California's current LEV I standards will remain in effect through MY2003.

ANALYSIS OF THE NATIONAL LOW EMISSION VEHICLE PROGRAM

The National LEV program adopts California's current LEV I standards for NMOG, $N0_x$, CO, PM, and HCHO. It also follows California's practice of allowing manufacturers to use fleet emission averages instead of certifying each vehicle according to the same stringency. The program focuses on controlling the formation of ground-level ozone pollution by requiring a 70% reduction in NMOG and a 50% reduction in NO_x. The standards require fewer reductions in CO primarily because ambient concentrations of this pollutant have been declining steadily since 1987. Gasoline-powered vehicles are not significant sources of PM, and the low emission standard for this pollutant is limited to diesel-powered vehicles. The standard for HCHO would reduce emissions of carcinogenic pollutants. However, vehicles certified under the National LEV program are likely to emit greater quantities of pollutants in actual use than the low emission standards indicate due to the use of less stringent emission standards for light trucks above 3,750 pounds and high sulfur levels in commercial gasoline. Sulfur in gasoline inhibits emission control systems, and gasoline used to certify vehicles according to the National LEV standards contains much less sulfur than commercial gasoline available outside of California. Specific requirements of the National LEV program, estimates of air quality benefits and costs, and the potential effects of gasoline sulfur levels and less stringent light truck standards on actual emissions are discussed below.

Certification Emission Standards

Similar to California's LEV I program, the National LEV program requires manufacturers to *certify* each vehicle according to one of four categories subject to increasingly stringent emission standards: federal Tier I, TLEVs, LEVs, or ULEVs. (Refer to Table I on page 7.) While the National LEV program does not include California's ZEV sales target, manufacturers also can certify their vehicles as ZEVs and receive credit for their air quality benefits. Relative to the NMOG standard, TLEVs are 50% cleaner than Tier I vehicles, LEVs are 70% cleaner, and ULEVs are 85% cleaner. Like the Tier I standards and California's

[15] For information on California's new LEV II standards, refer to the California Air Resources Board home page at [*http://www.arb.ca.gov*].

requirements, passenger cars and light trucks weighing up to 3,750 pounds are subject to stricter standards than heavier trucks weighing between 3,751 and 5,750 pounds. Trucks weighing more than 5,750 pounds are not regulated under the National LEV program and therefore remain subject to the less stringent Tier I standards. In addition, certified vehicles also must meet other low emission standards for the Supplemental Federal Test Procedure, which measures emissions during high speed acceleration, high speed driving, and air conditioning use.

Fleet Average Emission Standards

While certified vehicles are subject to low emission standards for a total of five pollutants, manufacturers will demonstrate compliance by averaging only the NMOG emissions of their vehicle fleets, which represent the most significant reductions under the program. Similar to the certification standards, passenger cars and light trucks up to 3,750 pounds are expected to meet stricter fleet emission averages than heavier trucks between 3,751 and 5,750 pounds. Vehicles sold in the Northeast in MY1999 and MY2000 will be expected to meet less stringent fleet emission averages than those that will become effective nationwide in MY2001. At this time, the fleet average emission standards will become as stringent as the LEV certification standards for NMOG measured to reflect an intermediate vehicle use of 50,000 miles. To preserve regional air quality benefits, manufacturers must calculate fleet emission averages for vehicles sold in the Northeast separately from those sold in other states.

Table 1. Comparison of Tier I and National LEV Standards

Vehicle Type/Weight	Intermediate Useful Life (grams/mile) [a]					Full Useful Life (grams/mile) [a]				
	NMHC/ NMOG [b]	CO	NO_x	PM [c]	HCHO	NMHC/ NMOG [b]	CO	NO_x	PM [c]	HCHO
Tier 1 Standards LDVs/LDTs (< 3,751 lbs.)	0.250	3.4	0.4	0.08	n/a	0.310	4.2	0.60	0.10	n/a
LDTs (3,751-5,750 lbs.)	0.320	4.4	0.7	0.08	n/a	0.400	5.5	0.97	0.10	n/a
National LEV Standards LDVs/LDTs (< 3,751 lbs.)										
Fleet Averages [d]										
MY1999 (Northeast)	0.148	n/a	n/a	n/a	n/a	n/a	n/a	n/a	n/a	n/a
MY2000 (Northeast)	0.095	n/a	n/a	n/a	n/a	n/a	n/a	n/a	n/a	n/a
MY2001(Nationwide)	0.075	n/a	n/a	n/a	n/a	n/a	n/a	n/a	n/a	n/a
Certification Standards										
TLEV	0.125	3.4	0.4	n/a	0.015	0.156	4.2	0.60	0.08	0.018
LEV	0.075	3.4	0.2	n/a	0.015	0.090	4.2	0.30	0.08	0.018
ULEV	0.040	1.7	0.2	n/a	0.008	0.055	2.1	0.30	0.04	0.011
LDTs (3,751-5,750 lbs.)										
Fleet Averages [d]										
MY1999 (Northeast)	0.190	n/a	n/a	n/a	n/a	n/a	n/a	n/a	n/a	n/a
MY2000 (Northeast)	0.124	n/a	n/a	n/a	n/a	n/a	n/a	n/a	n/a	n/a
MY2001(Nationwide)	0.100	n/a	n/a	n/a	n/a	n/a	n/a	n/a	n/a	n/a
Certification Standards										
TLEV	0.160	4.4	0.7	n/a	0.018	0.200	5.5	0.90	0.10	0.023
LEV	0.100	4.4	0.4	n/a	0.018	0.130	5.5	0.50	0.10	0.023
ULEV	0.050	2.2	0.4	n/a	0.009	0.070	2.8	0.50	0.05	0.013

[a] The intermediate useful life standards set a benchmark for the amount of pollutants that a vehicle is expected to emit after 5 years or 50,000 miles of vehicle use. The full useful life standards reflect the amount of pollutants that a vehicle is expected to emit after 10 years or 100,000 miles of vehicle use.

[b] Tier I vehicles are subject to the testing procedure for NMHC, whereas vehicles certified under the National LEV program are subject to the testing procedure for NMOG. Both pollutants are closely related hydrocarbon compounds that can combine with NO_x in the presence of sunlight to form ozone. The National LEV program adopted California's use of the testing procedure for NMOG because it more accurately measures the level of hydrocarbons in a vehicle's exhaust.

[c] The National LEV standards for PM apply only to diesel-fueled vehicles at 10 years or 100,000 miles.

[d] Each manufacturer has the flexibility to certify different models of vehicles according to the standards for TLEVs, LEVs, or ULEVs as long as the fleet average standard for NMOG emissions is met.

LDV = light duty (passenger) vehicle
TLEV = transitional low emission vehicle
ULEV = ultra low emission vehicle HCHO= formaldehyde
NMOG = non-methane organic gases
NO_x = nitrogen oxides LEV = low emission vehicle
NMHC = non-methane hydrocarbons
LDT = light duty truck CO = carbon monoxide
PM = particulate matter

Prepared by the Congressional Research Service based on data from the Environmental Protection Agency.

Trading Program

To provide additional flexibility in demonstrating compliance, manufacturers can trade NMOG credits to meet the fleet average emission standards. A manufacturer will generate credits in a model year if its fleet emissions average is lower than the standard, and will generate debits if its fleet emissions average exceeds the standard. Manufacturers with debits can purchase credits from others to achieve compliance. However, trading will be conducted separately in the Northeast to prevent manufacturers from using credits generated in states outside of the region with relatively clean air to achieve compliance in more heavily polluted northeastern states. Manufacturers also can save credits for use or sale in future model years. However, these credits will lose a portion of their value each year and will no longer be valid after the fourth year held. As an incentive to introduce cleaner vehicles nationwide sooner than the program requires, manufacturers are eligible for early reduction credits if they meet the fleet average emission standards in states outside of the Northeast prior to MY2001.

Estimated Air Quality Benefits

The main air quality benefit of the National LEV program would be reductions in the ozone precursors NMOG and NO_x. Other benefits would be reductions in PM, formaldehyde and benzene.[16] Based on the stringency of the certification standards, EPA estimates that by 2005 the National LEV program would reduce $N0_x$ emissions by 400 tons per day nationwide, NMOG by 279 tons, PM by 29 tons, benzene by 7 tons, and formaldehyde by 4 tons.[17] Whether the program is able to achieve these reductions will depend to a large extent on whether certified vehicles operate as cleanly in actual use as they do under test conditions. While the reductions estimated by EPA would help to reduce pollution from new vehicles, they are relatively small (less than 1%) compared to nationwide emissions from all sources combined.[18] On the other hand, such reductions could be more critical in urban areas where mobile sources typically account for a greater share of total pollution.

Estimated Costs

EPA analyzed the costs of the National LEV program based on cost estimates of vehicles sold in California. The California Air Resources Board estimates that the cost of an LEV in the state is $96 higher on average than a Tier I vehicle. EPA estimates a somewhat lower cost of $76 on average under the National LEV program due to four major reasons. First, advances in technology could reduce costs. Second, economies of scale also could lower costs since manufacturers are be able to produce a single fleet of vehicles subject to the same standards nationwide. Third, past cost estimates from the California Air Resources Board have proven higher than actual price differences. Fourth, industry historically has lowered vehicle prices in

---

16 Although there is no standard for benzene, EPA expects that the low emission technologies used to achieve the certification standards also would reduce emissions of this pollutant.

17 EPA. *Federal Register*. June 6, 1997. p. 31195-31196.

18 EPA. Office of Air Quality Planning and Standards. *1997 National Air Quality and Emissions Trends Report*. December 1998. p. 114-116. All sources combined emitted a daily average of 64,608 tons of NO_x, 52,641 tons of VOCs, and 8,526 tons of PM.

successive model years after the introduction of new technologies. Based on projections of vehicle sales nationwide, EPA estimates that the total cost of the National LEV program to consumers would be $950 million annually.[19]

Potential Effects of Gasoline Sulfur Levels on Actual Emissions

Vehicles certified under the National LEV program may not operate as cleanly in actual use as the certification standards suggest because high gasoline sulfur levels could impair the operation of emission controls. Sulfur occurs naturally in crude oil, and the amount remaining in gasoline varies widely among refineries. Gasoline sulfur levels average 340 ppm nationwide and can reach as high as 1,000 ppm.[20] Levels are lowest in California where a cap is currently in place that limits sulfur to a maximum of 80 ppm with a statewide average of 30 ppm. To meet emission standards, motor vehicles primarily depend on catalytic converters to reduce levels of CO, N0„ and VOCs from the exhaust. However, the sulfur in gasoline can react with the surface of the catalyst and inhibit the removal of other pollutants. The extent to which sulfur increases emissions depends mostly on the design of the catalyst, materials used, air to fuel mixture of the engine, and range of exhaust temperatures.

A discrepancy currently exists between sulfur levels in commercial gasoline and the amount in certification fuels. Under the National LEV program, manufacturers can certify their vehicles with the test fuel for California's LEV program or the federal test fuel for Tier I vehicles. Sulfur levels in California's test fuel are frequently below 30 ppm, and the amount in federal test fuel is typically less than 100 ppm, both of which are lower than average gasoline sulfur levels nationwide. Even in areas where the second phase of reformulated gasoline will be sold in 2000, sulfur levels will average as much as 150 ppm and continue to exceed test fuel amounts.[21] In effect, vehicles certified under the National LEV program may emit more pollution in actual use than the certification standards suggest, and therefore, produce fewer air quality benefits. However, the extent to which the air quality benefits of the program might be compromised is difficult to determine since different models of vehicles do not respond the same to high sulfur gasoline.

The vehicle manufacturing and oil refining industries have conducted studies to examine the effects of gasoline sulfur levels on emissions. They conclude that actual emissions depend on the amount of sulfur in gasoline, and NO_x emissions seem to be more affected by sulfur than other pollutants. However, sensitivity to sulfur varies among vehicles. EPA has combined the test results from two industry studies that indicate a wide range of effects among LEVs and ULEVs when operated on high sulfur gasoline. For example, emissions of NO_x from the 9 most sensitive vehicles increased by 251% when operated on gasoline with a national sulfur average of 3 40 ppm, whereas NO_x emissions from the 9 least sensitive vehicles increased by only 61%. These studies also tested the effects of sulfur on Tier I vehicles and observed a lower increase in NO_x emissions of 13.6%, which indicates that the

[19] EPA. *Federal Register.* June 6, 1997. p. 31197.

[20] EPA. Office of Mobile Sources. *Staff Paper on Gasoline Sulfur Issues.* May 1998. p. 3.

[21] Reformulated gasoline (RFG) bums more cleanly and results in fewer emissions of ozone- forming and toxic compounds. Under the Clean Air Act, 10 metropolitan areas with serious air quality problems require the sale of RFG, and 13 additional areas voluntarily require the sale of RFG. The majority of these areas are located in the Northeast.

technologies used to achieve the LEV and ULEV standards are significantly more sensitive than those typically used in Tier I vehicles.[22]

Potential Effects of Light Truck Standards on Actual Emissions

Like the Tier I standards under the Clean Air Act and California's current LEV I standards, the National LEV program continues the practice of allowing light trucks over 3,750 pounds to meet less stringent emission standards. Congress initially established more lenient standards for these heavier trucks because their capacity for carrying large loads compromised emissions performance and they composed a smaller portion of the vehicle fleet, which made less of an impact on air quality. However, light trucks now account for a much larger share of total vehicle sales. According to industry sales data, light trucks accounted for about 20% of total vehicle sales in 1980.[23] By 1998, the market share of these vehicles had increased to nearly 48%.[24] The sharp rise in sales is mostly due to the heightened popularity of sport utility vehicles and minivans. Sport utility vehicles alone accounted for about 18% of total vehicle sales in 1998, while minivans accounted for roughly 8%.[25] However, the historical change in the market share of light trucks somewhat underestimates the actual growth in sales. While the number of passenger cars sold in the United States has decreased since 1980, the number of light trucks sold has more than tripled from 2.2 million in 1980 to 7.4 million in 1998. Consequently, the contribution of light trucks to total emissions from motor vehicles has increased as well.[26]

Based on the number and weight of vehicles certified to meet the federal Tier I standards, EPA estimates that about 37% of all vehicles sold outside of California in 1998, including light trucks, sport utility vehicles, and minivans, weigh in excess of 3,750 pounds.[27] These heavier vehicles are subject to less stringent emission standards even though they are used primarily for personal transportation rather than for carrying heavy loads, which the more lenient standards were initially designed to accommodate. The increasing market share of heavier light trucks, sport utility vehicles, and minivans could alter the composition of the vehicle fleet significantly enough to increase total emissions and offset the sharper reductions from lighter cars and trucks. As a result, overall reductions in vehicle emissions under the National LEV program might be less promising.

---

[22] EPA. Office of Mobile Sources. *Staff Paper on Gasoline Sulfur Issues.* May 1998. p. II. Test results compiled from two studies conducted by the Coordinating Research Council (CRC), representing the oil industry, and by the American Automobile Manufacturers Association and the Association of International Automobile Manufacturers.

[23] Automotive News. *The 100-Year Almanac and 1996 Market Data Book.* p, 120.

[24] Automotive News. *1999 Market Data Book.* p. 3 5.

[25] Ibid., p. 33.

[26] For further discussion, refer to CRS Report RS20298, *Sport Utility Vehicles, Mini-Vans and Light Trucks: An Overview of Fuel Economy and Emission Standards,* by Brent D. Yacobucci.

[27] EPA. Office of Mobile Sources. *Tier 2/Sulfur Draft Regulatory Impact Analysis.* April 1999. p. V-24-25. The full text of the document is available electronically from EPA at [*http://www.epa.gov/oms/tr2home.htm*].

ANALYSIS OF EPA'S TIER 2 PROPOSAL

While states and vehicle manufacturers have voluntarily agreed to participate in the National LEV program, compliance with EPA's Tier 2 proposal would be mandatory if finalized. EPA based its proposal on California's LEV II program to ensure that vehicle manufacturers remain subject to the same requirements nationwide. Following California's example, the proposed Tier 2 standards would address the issue of rising light truck emissions by regulating all new passenger cars and light trucks up to 8,500 pounds according to the same emission standards. The proposed standards would be phased in gradually between MY2004 and MY2009. All new passenger cars and light trucks up to 6,000 pounds would be required to achieve full compliance by MY2007, but heavier trucks between 6,001 and 8,500 pounds would not be required to meet the same standards until two years later in MY2009. EPA is also proposing to follow California's example of regulating gasoline sulfur levels to improve the operation of emission control technologies. The Tier 2 proposal would limit gasoline sulfur levels to an average of 30 ppm nationwide beginning in 2004, roughly 90% less than the current national average of 340 ppm.

Like the National LEV program, the proposed Tier 2 standards would limit tailpipe emissions of NMOG, $N0_x$, CO, PM, and HCHO. The proposed standards would require the greatest reductions in emissions of NMOG and NO_x to help control the formation of ground-level ozone pollution. Relative to the Tier I standards, the fleet average emissions standard for NO_x would require vehicle manufacturers to reduce overall emissions by 88% to 95%. The proposed standards also would require at least an 80% reduction in PM emissions. Unlike the National LEV program, the proposed PM standard would apply to both gasoline and diesel-powered vehicles. Limits on emissions of CO and HCHO would be similar to those under the National LEV program. Manufacturers would have the flexibility to select from among several different sets of emission standards instead of certifying each vehicle according to the same stringency. While the certification standards would limit emissions of all five regulated pollutants, EPA would determine compliance by averaging the NO_x emissions of a manufacturer's vehicle fleet, which would represent the largest reductions under the program. The use of fleet emission averages would allow manufacturers broad flexibility in selecting which set of standards applies to different models of vehicles, while still meeting the air quality objective of lowering overall N0,, emissions to a specific level. Key elements of EPA's Tier 2 proposal, estimates of air quality benefits and costs, responses from industry and environmental organizations, and potential legal challenges that some opponents might raise are discussed below.

Certification Emission Standards

Manufacturers would be able to select from among several different sets (or "bins") of emission standards when certifying their vehicles. (Refer to Table 2 on page 13.) Each bin would include separate limits on emissions measured to reflect an intermediate and full useful life of vehicle operation. The benchmark for a vehicle's intermediate useful life would continue to be 50,000 miles, as the Tier I and National LEV standards require. However, EPA's proposal would increase the full useful life benchmark from 100,000 to 120,000 miles to match the performance requirement adopted under California's LEV II program. The *least* stringent emissions bin would serve as a cap on the maximum amount of pollution that each vehicle would be allowed to emit and is based on the Tier 2 guidelines specified in Section 202(i) of the Clean Air Act. As such, it would allow a certified vehicle to emit .20 grams per mile of N0„ nearly three times as much as the fleet average emission standard of .07 grams per mile that EPA is proposing. The *most* stringent bin would be reserved solely for certifying ZEVs, which would require that emissions of all five regulated pollutants be reduced to unmeasurable levels. While EPA's proposal does not include a ZEV sales mandate, the agency included a bin for such vehicles to allow manufacturers that make them to receive credit for their air quality benefits when calculating their fleet emission averages. In addition to the certification standards, each vehicle also would be required to meet stricter limits on evaporative emissions and tougher requirements under the Supplemental Federal Test Procedure, which measures emissions during high speed acceleration, high speed driving, and air conditioning use. During the phase-in period beginning in MY2004, vehicles not certified according to the Tier 2 standards would be subject to a less stringent set of interim standards which are based on the National LEV program.

Fleet Average Emission Standards

While each vehicle would be subject to limits on emissions of all five regulated pollutants, EPA would determine compliance by averaging the N0„ emissions of each manufacturer's vehicle fleet. The fleet average emissions standard would limit N0„ emissions to .07 grams per mile measured at a full useful life of 120,000 miles. All passenger cars and light trucks up to 6,000 pounds would be required to meet this standard in MY2007, and all heavier trucks between 6,001 and 8,500 pounds would become subject to it two years later in MY2009. During the phase-in period, manufacturers also would be subject to less stringent fleet emission averages for vehicles certified according to the interim standards. The interim fleet average emissions standard for passenger cars and light trucks up to 6,000 pounds would be .30 grams per mile, but heavier trucks between 6,001 and 8,500 pounds would be subject to a stricter interim fleet average emissions standard of .20 grams per mile due to the more lenient phase-in schedule for these heavier vehicles. While the National LEV program and the state of California require each manufacturer to average its fleet emissions of NMOG to determine compliance, EPA expects that its proposal to use fleet average emission standards for NO_x would result in comparable reductions of NMOG as well.

Table 2. Comparison of Tier I and Proposed Tier 2 Standards

Vehicle Type/Weight	Intermediate Useful Life (grams/mile) [a]					Full Useful Life (grams/mile) [a]				
	NMHC/ NMOG [b]	**CO**	**NO_x**	**PM**	**HCHO**	**NMHC/ NMOG** [b]	**CO**	**NO_x**	**PM**	**HCHO**
Tier 1 Standards LDVs/LDTs (< 3,751 lbs.)	0.250	3.4	0.40	0.08	n/a	0.310	4.2	0.60	0.10	n/a
LDTs (3,751-5,750 lbs.)	0.320	4.4	0.7	0.08	n/a	0.400	5.5	0.97	0.10	n/a
LDTs (>5,750 lbs.)	0.390	5.0	1.10	n/a	n/a	0.560	7.3	1.53	0.12	n/a
Tier 2 Interim Standards [c]										
LDVs/LDTs (<6,001 lbs.)										
Fleet Average [d]	n/a	n/a	n/a	n/a	n/a	n/a	n/a	0.30	n/a	n/a
Bin 5	0.125	3.4	0.40	n/a	0.015	0.156	4.2	0.60	0.06	0.018
Bin 4	0.075	3.4	0.20	n/a	0.015	0.90	4.2	0.30	0.06	0.018
Bin 3	0.40	1.7	0.20	n/a	0.008	0.055	2.1	0.30	0.04	0.011
Bin 2	0.075	3.4	0.05	n/a	0.015	0.090	4.2	0.07	0.01	0.018
Bin 1	n/a	n/a	n/a	n/a	n/a	0.000	0.0	0.00	0.00	0.000
HLDTs (6,001-8,500 lbs.)										
Fleet Average [d]	n/a	n/a	n/a	n/a	n/a	n/a	n/a	0.20	n/a	n/a
Bin 5	0.160	3.4	0.40	n/a	0.015	0.230	4.2	0.60	0.06	0.018
Bin 4	0.140	3.4	0.20	n/a	0.015	0.180	4.2	0.30	0.06	0.018
Bin 3	0.125	3.4	0.14	n/a	0.015	0.156	4.2	0.20	0.02	0.018
Bin 2	0.075	3.4	0.05	n/a	0.015	0.090	4.2	0.07	0.01	0.018
Bin 1	n/a	n/a	n/a	n/a	n/a	0.000	0.0	0.00	0.00	0.000

Vehicle Type/Weight	Intermediate Useful Life (grams/mile) [a]					Full Useful Life (grams/mile) [a]				
	NMHC/ NMOG [b]	**CO**	**NO_x**	**PM**	**HCHO**	**NMHC/ NMOG** [b]	**CO**	**NO_x**	**PM**	**HCHO**
Tier 2 Basic Standards										
LDVs/LDTs (<6,001 lbs.)										
HLDTs (6,001-8,500 lbs.)										
Fleet Average [d]	n/a	n/a	n/a	n/a	n/a	n/a	n/a	0.07	n/a	n/a
Bin 7	0.100	3.4	0.14	n/a	0.015	0.125	4.2	0.20	0.02	0.018
Bin 6	0.075	3.4	0.11	n/a	0.015	0.090	4.2	0.15	0.02	0.018
Bin 5	0.075	3.4	0.05	n/a	0.015	0.090	4.2	0.07	0.01	0.018
Bin 4	0.040	1.7	0.05	n/a	0.008	0.055	2.1	0.07	0.01	0.011
Bin 3	n/a	n/a	n/a	n/a	n/a	0.070	2.1	0.04	0.01	0.011
Bin 2	n/a	n/a	n/a	n/a	n/a	0.010	2.1	0.02	0.01	0.004
Bin 1	n/a	n/a	n/a	n/a	n/a	0.000	0.0	0.00	0.00	0.000

[a] The intermediate useful life standards set a benchmark for the amount of pollutants that a vehicle is expected to emit after being driven for 50,000 miles. The Tier I full useful life standards reflect the amount of pollutants that a vehicle is expected to emit after 100,000 miles of vehicle use. EPA's Tier 2 proposal would increase the full useful life standards to 120,000 miles.

[b] Tier 1 vehicles are subject to the testing procedure for NMHC, but Tier 2 vehicles would be subject to the testing procedure for NMOG. Both pollutants are closely related hydrocarbon compounds that can combine with NO_x in the presence of sunlight to form ozone. EPA is proposing to use the NMOG testing procedure under Tier 2 because it more accurately measures hydrocarbon levels.

[c] During the phase-in period that begins in MY2004, vehicles not certified according to the Tier 2 basic standards would be subject to the interim standards. The phase-in period would end for LDVs and LDTs in MY2007 and for HLDTs in MY2009.

[d] Each manufacturer would have the flexibility to select any set, or "bin", of emission standards when certifying different models of vehicles, as long as the fleet average standard for NO_x emissions is met.

HLDT = heavy light duty truck NMHC = non-methane hydrocarbons NO_x = nitrogen oxides
LDV = light duty (passenger) vehicle NMOG = non-methane organic gases PM = particulate matter
LDT = light duty truck CO = carbon monoxide HCHO = formaldehyde

Prepared by the Congressional Research Service based on data from the Environmental Protection Agency.

Phase-in Schedule

Passenger cars and light trucks up to 6,000 pounds would be subject to an earlier phase-in schedule than heavier trucks between 6,001 and 8,500 pounds. Vehicle manufacturers would be required to certify at least 25% of their passenger cars and light trucks up to 6,000 pounds according to the Tier 2 standards in MY2004, at least 50% in MY2005, at least 75% in MY2006, and 100% in MY2007. Vehicles in this weight class that are not certified according to the Tier 2 standards between MY2004 and MY2006 would be subject to the interim standards. Manufacturers would not be required to certify any of their light trucks between 6,001 and 8,500 pounds according to the Tier 2 standards until MY2008. However, they would be required to certify at least 25% of these heavier vehicles according to the interim standards in MY2004, at least 50% in MY2005, at least 75% in MY2006, and 100% in MY2007. Trucks in this heavier weight class that are not certified according to the interim standards between MY2004 and MY2006 would remain subject to the current Tier I standards under the Clean Air Act. Beginning in MY2008, manufacturers would be required to certify at least 50% of their light trucks weighing between 6,001 and 8,500 pounds according to the Tier 2 standards with the remaining 50% still being subject to the interim standards. The phase-in period would end in MY2009 when all new passenger cars and light trucks weighing up to 8,500 pounds would be subject to the Tier 2 standards, at which time the interim standards would no longer apply.

Vehicle Emissions Trading Program

Similar to the National LEV program, EPA's proposal would provide additional flexibility by allowing vehicle manufacturers to bank and trade $N0_x$ credits that could be used to demonstrate compliance. Manufacturers would earn credits for reducing their overall $N0_x$ emissions below the fleet average emissions standard, and manufacturers having difficulty in controlling their fleet emissions could purchase such credits to achieve compliance. EPA also would allow manufacturers to bank any earned or purchased credits for use or sale in the future. While credits banked under the National LEV program lose their value after 4 years, EPA's Tier 2 proposal would not restrict the number of years that NO_x credits could be banked prior to their use. If a manufacturer is unable to meet the fleet average emissions standard after using all of the credits that it holds, an emissions deficit could be carried into the next model year. However, the manufacturer would have to earn or purchase enough credits during that model year to offset the deficit or face enforcement actions. During the phase-in period, NO_x credits earned for reducing overall emissions below the less stringent interim fleet average emission standards would be traded separately from those generated for controlling emissions beyond the stricter fleet average emissions standard for Tier 2 vehicles. Requiring such credits to be traded separately would prevent manufacturers from making large amounts of easier reductions under the interim standards to comply with the tougher Tier 2 requirements.

National Limits on Gasoline Sulfur Levels

EPA based its proposal to limit gasoline sulfur levels on fuel formula requirements that have been in effect in California since 1996. Each oil refinery would be required to reduce sulfur levels in gasoline to an average of 30 ppm beginning in 2004, a reduction of over 90% below the current national average of 340 ppm. Refineries would also be subject to a less stringent cap on the amount of sulfur in each gallon of gasoline. The cap would be phased in gradually and limit sulfur levels to 300 ppm per gallon in 2004, 180 ppm per gallon in 2005, and 80 ppm per gallon in 2006 and thereafter. However, small refineries would be subject to more lenient gasoline sulfur limits between 2004 and 2007 and would not be required to comply with the 30 ppm average standard and 80 ppm cap until 2008.[28] Hardship extensions also would be available to small refineries that could extend the compliance deadline to 2010. EPA expects that a more lenient compliance schedule for small refineries would help to ease the regulatory burden on such businesses without significantly affecting the emission reduction potential of Tier 2 vehicles, since small refineries produce only about 2.5% of all gasoline in the United States.

Gasoline Sulfur Trading Program

Similar to the trading program for vehicle manufacturers, the gasoline sulfur trading program would allow the oil refining industry the flexibility to bank or trade credits to meet the national limit on average gasoline sulfur levels. Oil refiners would generate credits for reducing their average sulfur levels below the 30 ppm standard. Such credits could be used in the year in which they were generated, banked for future use, or sold to other refiners having difficulty in demonstrating compliance. However, banked credits would have a limited useful life of 5 years from the time that they were generated or purchased, whereas the trading program for vehicle manufacturers would allow unlimited use of banked credits. While refiners could use credits to comply with the average sulfur standard of 30 ppm, they would still be subject to the per gallon cap on sulfur levels of 300 ppm in 2004, 180 ppm in 2005, and 80 ppm in 2006 and thereafter. Like the trading program for vehicle manufacturers, the gasoline sulfur trading program would allow an oil refiner who is unable to demonstrate compliance after using all of the credits that it holds to carry the resulting deficit into the next year. The refiner would then have to earn or purchase enough credits during that year to offset the deficit or face enforcement actions. Additionally, refiners could earn early credits between 2000 and 2003 for reducing average sulfur levels to 150 ppm or less and could use these early credits to help them achieve compliance when the 30 ppm standard becomes effective in 2004. However, oil refiners would be required to use early credits by 2007, after which time such credits would entirely lose their value.

---

[28] EPA uses the Small Business Administration's definition of a small petroleum refining business as one that does not have more than 1,500 employees. Approximately 17 of the 160 refineries in the United States meet this definition, and EPA reports that 9 of these 17 refineries already produce gasoline with a sulfur content of 90 ppm or less. The oil refining industry argues that a different definition should be used to allow additional refiners to qualify for the more lenient standards. For a discussion of this issue, refer to CRS Report RS20163, *Sulfur in Gasoline,* by Stephen Thompson and James McCarthy.

Estimated Air Quality Benefits

Like the National LEV program, the primary air quality benefit of EPA's Tier 2 proposal would be reductions in the ozone precursors NO_x and NMOG. Other benefits would include reductions in PM, CO, and the carcinogenic pollutants formaldehyde and benzene.[29] Based on the stringency of the Tier 2 standards and the projected number of vehicle miles traveled in future years, EPA estimates that vehicular NO_x emissions would decline by 25% from 3.2 to 2.4 million tons in 2007 when passenger cars and light trucks up to 6,000 pounds would be required to achieve full compliance. At this time, EPA predicts that such reductions would lower the number of metropolitan areas expected to exceed the 1-hour NAAQS for ozone from 8 to 4 and would reduce those expected to violate the revised 8-hour NAAQS for ozone from 28 to 25.[30] In 2010, the Tier 2 standards are expected to assist 7 additional areas in meeting the revised 8-hour standard. By 2020, EPA projects that vehicular $N0_x$ emissions would decline by 68% from 3.2 to 1.0 million tons once Tier 2 vehicles represent the majority of the on-road vehicle fleet.[31]

While EPA expects that the above reductions in NO_x emissions would lower the overall amount of ground-level ozone pollution, the agency acknowledges that a few locations in the centers of large metropolitan areas could actually experience an increase in air quality problems. The atmospheric chemistry of these areas causes a reverse effect which tends to result in higher ozone levels as concentrations of NO_x fall. In such areas, reducing VOCs, rather than NO_x, is the critical factor in controlling ground-level ozone pollution. However, EPA predicts that this effect would begin to diminish as these areas implement planned measures to reduce VOCs, since reducing both VOCs and NO_x in significant amounts at the same time ultimately results in progressively lower ozone levels. EPA may explore this issue further in its final rule and might present sufficient data to identify which metropolitan areas and how many people could experience higher ground-level ozone pollution as a result of NO_x reductions under the Tier 2 standards.

In addition to controlling overall ground-level ozone pollution, EPA also expects that large reductions in emissions of particulate matter would occur as a result of the Tier 2 standards. Vehicular emissions of particulate matter are expected to decline by 60% in 2007 from 39,209 to 15,830 tons and by 63% in 2020 from 51,102 to 19,071 tons.[32] Reductions in particulate matter would likely be high from the start instead of rising substantially over time because all vehicles, and not just those certified to meet the Tier 2 standards, should emit less particulate matter when operated on lower sulfur gasoline. However, the reductions estimated by EPA would have limited effects on overall air quality because they are relatively small (around 1%) compared to total emissions of particulate matter from all sources combined.

---

[29] EPA expects that emission control measures used to meet the Tier 2 formaldehyde standard would also result in lower levels of benzene emissions, even though there would be no standard for this pollutant.

[30] EPA revised the NAAQS for ozone in July 1997. The revised standard for ozone is .08 ppm measured over 8 hours. The previous standard was .12 ppm measured over one hour. States that have not yet met the 1-hour standard are required to do so before being subject to the revised 8-hour standard. For further discussion, refer to CRS Report 97-8, *Air Quality: Background and Analysis of EPA's 1997 Ozone and Particulate Matter Standards,* by John Blodgett, Larry Parker, and James McCarthy.

[31] EPA. *Federal Register.* May 13, 1999. p. 26016-26018.

[32] Ibid., p. 26020.

Estimated Costs

EPA estimates that the total costs of meeting the Tier 2 vehicle emission standards and national limits on gasoline sulfur levels would be approximately $3.5 billion annually once these requirements are fully phased in.[33] The total annual cost of $3.5 billion is expected to be nearly equal to or exceed the monetized value of public health benefits, which may range from a low estimate of $3.2 billion to as much as $19.5 billion annually.[34] EPA estimates that the vehicle manufacturing industry would likely need to spend as much as $5 million in research and development for each vehicle line certified to meet the stricter Tier 2 emission standards.[35] However, the actual costs of research and development may be lower than this amount if innovations in one vehicle line could be applied to others. Additionally, EPA estimates that the oil refining industry would likely need to spend $4.7 billion in initial capital investments to install the technologies that would be necessary to reduce average gasoline sulfur levels to 30 ppm. The annual cost of operating these technologies could be as high as $1.5 billion annually.[36]

EPA estimates that the costs of complying with a national limit on average gasoline sulfur levels of 30 ppm would be 1.7 cents per gallon during the first year of compliance in 2004, but would decline over time to a low of 1.4 cents per gallon in 2015.[37] However, these estimates are based upon the assumption that recent advancements in desulfurization technologies would become widely available to most refiners in the near future. Based on currently available technologies, the oil refining industry has estimated a higher cost of 5 to 6 cents per gallon to meet an average sulfur content of 30 ppm nationwide.[38] More recent industry estimates predict a lower cost of about 2.5 cents per gallon if EPA were to adopt a less stringent sulfur limit of 40 ppm and restrict its applicability to the eastern portion of the country where regional ozone problems are more severe.[39]

Increases in the purchase price of vehicles certified to meet the Tier 2 standards would likely be higher for heavier trucks since these vehicles generally need more extensive modifications to control emissions. For example, EPA estimates that the purchase price of light trucks over 5,750 pounds would initially increase by $266 in the first year of production and then fall to an increase of $209 during the 6~ year and thereafter.[40] In comparison, the price of passenger cars and light trucks up to 3,750 pounds is expected to increase by roughly $70 during the first year of production and then decline to an increase of around $45 during the 6* year and thereafter. EPA predicts that increases in the purchase price of Tier 2 vehicles would experience a downward trend after the first few years of production due to three major factors. First, production efficiencies typically improve over time. Second, more efficient and less costly technologies to control pollution may become available as a result of ongoing research. Third, manufacturers may be able to recover capital investments in more advanced emission control systems within a few years.

---

[33] Ibid., p. 26072.

[34] Ibid., p. 26078-26079.

[35] Ibid., p. 26070.

[36] Ibid., p. 26071.

[37] Ibid.

[38] *Lundberg Letter,* Vol. 25, no. 10. May 27, 1998. p. 1.

[39] MathPro Inc., Bethesda, Maryland. *Costs of Meeting a 40ppm Sulfur Content Standard for Gasoline in PADDS 1-3, via Mobil and CD Tech Desulfurization Processes.* February 26,1999. p. 2.

[40] EPA. *Federal Register.* May 13, 1999. p. 26071.

Responses from Industry and Environmental Organizations

The vehicle manufacturing industry, the oil refining industry, and some environmental organizations have expressed numerous concerns about EPA's Tier 2 proposal. The 106th Congress has held four hearings which focused on the potential regulation of gasoline sulfur levels. The Subcommittee on Clean Air of the Senate Committee on Environment and Public Works conducted hearings on May 18, 1999, May 19, 1999, and July 21, 1999. The Subcommittee on Energy and Environment of the House Committee on Science also conducted a hearing on July 29, 1999. In addition to testimony received by the 106th Congress, EPA held a series of public field hearings on the proposed Tier 2 standards from June 9-17, 1999 in Philadelphia, Atlanta, Denver, and Cleveland, and received public comments until August 2, 1999.[41] Major views that have been expressed by the vehicle manufacturing industry, the oil refining industry, and some environmental organizations are summarized below.

Vehicle Manufacturing Industry

Vehicle manufacturers generally support stricter Tier 2 emission standards if gasoline sulfur levels are regulated accordingly. However, the Alliance of Automobile Manufacturers, as well as individual manufacturers, have voiced at least three concerns over EPA's proposal. First, gasoline sulfur levels lower than 30 ppm might be necessary to ensure that advanced emission control technologies function properly. For example, near zero to 5 ppm sulfur levels would be necessary for direct-injection engine technologies to be maximally effective in reducing emissions. Second, additional and less stringent emission bins might be necessary for the heaviest trucks to meet the certification standards without impairing vehicle performance. Third, requiring heavier trucks between 6,001 and 8,500 pounds to meet the same emission standards as lighter vehicles by MY2009 may be too costly. A later compliance schedule for heavier trucks might be necessary to allow additional time for more affordable and effective emission control technologies to become widely available.

Oil Refining Industry

While oil refiners generally agree that sulfur does impair the performance of emission control technologies, the National Petrochemical and Refiners Association and numerous individual refiners have expressed at least three major reservations about EPA's Tier 2 proposal. First, the compliance schedule of 2004 to 2006 might not allow sufficient time for promising developments in desulfurization technologies to become commercially available to most refineries. Second, they argue that gasoline sulfur levels could be regulated as effectively and at a lower cost on a regional basis, primarily in the eastern portion of the country where air quality problems are more severe. Third, the sulfur trading program may provide little regulatory relief because not enough time would be allowed to generate enough credits prior to the time at which the standards would take effect in 2004. In light of these concerns, the oil refining industry warns that small refineries could be forced out of business and gasoline prices and supplies could be affected if EPA does not adopt a later phase-in schedule and offer the industry more flexibility.

[41] Transcripts from each public hearing are available electronically from EPA at [*http://www.epa.oms/tr2home.htm*].

Environmental Organizations

The environmental community has generally applauded EPA for proposing more stringent emission standards for motor vehicles. However, some environmental organizations, such as the Sierra Club, the U.S. Public Interest Research Group, and others, have expressed numerous concerns about EPA's Tier 2 proposal. First, the later compliance deadline of MY2009 for light trucks over 6,000 pounds would delay some of the emission reductions and reduce the immediate air quality benefits of the stricter standards. Second, the upper weight limit of 8,500 pounds would exclude the largest sport utility vehicles from regulation under the Tier 2 standards. While these vehicles currently represent a relatively small share of total vehicle sales, their rising popularity and use as passenger vehicles could result in higher pollution levels. Further, the proposed standards may encourage some manufacturers to increase the weight of certain vehicles that are just under 8,500 pounds so that these vehicles escape regulation. Third, the two top emission bins are not strict enough and would permit certain vehicles to pollute more excessively than California's LEV II program would allow. Fourth, they argue that the phase-in schedule for the sulfur standards would be too late because lower sulfur levels would be necessary before 2004 to maximize the air quality benefits of the National LEV program when it becomes effective nationwide in MY2001. Fifth, permitting vehicle manufacturers and oil refiners to carry compliance deficits into the following year would allow emissions to exceed the stricter standards in the short-term.

Potential Legal Challenges

As discussed above, numerous aspects of EPA's proposal could motivate some opponents to legally challenge the Tier 2 standards in court if the agency's final rule does not satisfy their concerns. However, a court ruling on the revised NAAQS for ozone and particulate matter might have a more direct bearing on potential litigation. A U.S. Court of Appeals remanded the revised NAAQS for ozone and particulate matter on May 14, 1999 partly because it viewed EPA's promulgation of the stricter standards under the Clean Air Act as an unconstitutional delegation of congressional authority to an executive agency.[42] The court also remanded the revised NAAQS because it felt that EPA did not adequately justify why the specific numerical standards that the agency promulgated were necessary to protect public health. EPA primarily based its justification of the proposed Tier 2 standards on the expectation that further reductions in emissions from motor vehicles would be necessary for states to attain and maintain both the 1-hour and the revised 8-hour NAAQS for ozone as well as the revised standards for particulate matter. Since the court found EPA's justification of the revised NAAQS to be insufficient, some opponents of the Tier 2 proposal have taken the opportunity to question the need for further emission reductions from motor vehicles to assist states in complying with more stringent NAAQS that may ultimately be overturned or significantly modified.

To deter potential legal challenges to the Tier 2 standards once they are finalized, EPA issued a supplemental notice on June 30, 1999 that clarifies the need for stricter vehicle

[42] D.C. Circuit Court of Appeals. *American Trucking Associations v. EPA.* No. 97-1440. May 14, 1999. For a summary of the Court of Appeals ruling and an explanation of the legal doctrine of nondelegation, refer to CRS Report RS20228, *The D.C. Circuit Remands the Ozone and Particulate Matter Clean Air Standards: American Trucking Associations v. EPA,* by Robert Meltz and James McCarthy.

emission standards and national limits on gasoline sulfur levels to attain and maintain the 1-hour NAAQS for ozone.[43] To further substantiate its claim, EPA presented an alternative analysis of air quality modeling data which predicts that 17 metropolitan areas outside of California will likely exceed the 1-hour NAAQS for ozone by 2007. As discussed earlier, the agency originally predicted that a total of 8 metropolitan areas would be in violation of the 1-hour ozone standard without stricter Tier 2 controls in place. Despite EPA's claims that further emission reductions from motor vehicles would be necessary to meet the 1-hour ozone standard, some opponents may still challenge the agency's authority to promulgate stricter controls based on the doctrine of unconstitutional delegation invoked in the Court of Appeals ruling on the revised NAAQS. However, if the court ultimately upholds the revised NAAQS, such a decision would affirm EPA's authority to promulgate stricter air quality standards, which could deter opponents of the Tier 2 standards from legally challenging them.

CONGRESSIONAL INTEREST AND LEGISLATIVE ACTIVITY

Implementation of the National LEV program began in MY 1999. A total of 23 automobile manufacturers are voluntarily producing vehicles for sale in 8 northeastern states and the District of Columbia, which are certified to meet low emission standards that are more stringent than the current Tier I requirements under the Clean Air Act. Before the National LEV standards become nationwide in MY2001, oversight of the program is a possibility in the 106th Congress to determine whether it is functioning as the automobile industry, the northeastern states, and EPA intended. Some of the issues that could possibly be examined during oversight include whether the use of fleet emission averages and tradeable credits offers automobile manufacturers sufficient flexibility to demonstrate compliance, the extent to which vehicle prices have increased as a result of using more expensive pollution control technologies, and whether stricter emission standards are producing the amount of air quality benefits that EPA predicted would occur under the program.

Congressional interest in EPA's Tier 2 proposal has focused primarily on the potential regulation of gasoline sulfur levels, rather than on the stricter emission standards that would apply to new motor vehicles. Some Members of Congress have expressed opposition to the proposed limits on gasoline sulfur levels due to the effects that they could possibly have on small refineries and gasoline supplies and prices. Senator Inhofe, Chairman of the Senate Environment Subcommittee on Clean Air, has suggested that he might introduce legislation to modify or revoke the Tier 2 sulfur controls if EPA finalizes them as proposed. However, other Members of Congress have indicated support for EPA's Tier 2 proposal, and legislation has been introduced that would regulate the sulfur content of gasoline nationwide. To date, legislation has not been introduced in the 106th Congress that would address emission standards for motor vehicles. Considering the controversial nature of the proposed Tier 2 standards, congressional oversight is a possibility after EPA issues a final rule.

Earlier this year, two bills were introduced in the House and Senate that would require the content of sulfur in commercial gasoline to be reduced to levels that are similar to what EPA is proposing. Senator Moynihan introduced the Clean Gasoline Act of 1999 (S. 171) on January 19, 1999, and Representative Kildee introduced a companion bill (H.R. 888) in the

[43] EPA. *Federal Register.* June 30, 1999. p. 35112-35119.

House on March 1, 1999. Beginning 4 years after enactment, each bill would limit gasoline sulfur levels to 40 ppm per gallon nationwide, or alternatively to 80 ppm per gallon as long as the average sulfur content of a refiner's annual production is less than 30 ppm. Six years after enactment, each bill also would require EPA to submit a report to Congress on the extent to which low sulfur gasoline has helped to improve urban and regional air quality. If EPA's findings were to indicate that further reductions in sulfur would result in significant air quality benefits or would enable advanced technologies to control emissions more effectively, the agency would be granted the authority to regulate gasoline sulfur levels more stringently than the amounts stipulated in the bill.

CONCLUSION

The automobile industry has been complying with low emission standards in California for the past few years. During this time, manufacturers have demonstrated that the technology to produce cleaner vehicles is currently available. However, the ability of this technology to reduce emissions in actual use depends on how sensitive a vehicle is to the amount of sulfur in gasoline. At this time, California is the only state that requires the sale of low sulfur gasoline to ensure that low emission vehicles operate as cleanly in actual use as they do under test conditions. Consequently, efforts under the National LEV program to reduce vehicle emissions with current technologies in areas where cleaner fuels are not available could prove extremely difficult In addition, the rising market share of heavier trucks, sport utility vehicles, and minivans could result in expanding the emissions inventory of the nation's on-road vehicle fleet and offset some of the sharper reductions in emissions from passenger cars and lighter weight trucks certified under the National LEV program. While Congress has not pursued an oversight role thus far, it may do so as the implementation of the program expands beyond the Northeast to include the rest of the country in MY2001. Through its Tier 2 proposal, EPA is attempting to address the regulatory issues of gasoline sulfur levels and light truck emissions which were left unresolved under the National LEV program. Whether the Tier 2 standards would be economically feasible and attainable within the time frame that EPA has proposed depends on numerous factors, including the extent to which advanced pollution control technologies are able to reduce emissions from the heaviest trucks to the same degree as lighter vehicles and the commercial availability of recent advancements in technologies to control gasoline sulfur levels. The number of areas that will need the amount of emission reductions which EPA is proposing will depend in part on whether the revised NAAQS for ozone and particulate matter are upheld, modified, or overturned. The vehicle manufacturing industry, the oil refining industry, and some environmental organizations have expressed numerous reservations about the Tier 2 proposal. If their concerns are not adequately addressed in EPA's final rule, some opponents might legally challenge the Tier 2 standards in court, which possibly could result in delaying their implementation or perhaps even overturn them. In Congress, there is both support for and opposition to the Tier 2 proposal. While legislation has been introduced to regulate the amount of sulfur in commercial gasoline according to levels that are similar to what EPA is proposing, legislation to scale back the proposed Tier 2 controls or to address emission standards for motor vehicles has not been introduced to date.

Chapter 12

AIR QUALITY AND THE NEW OZONE NAAQS: THE OTAG PROCESS*

Larry B. Parker and John E. Blodgett

INTRODUCTION

The Administration has announced that the recommendations of the Ozone Transport Assessment Group (OTAG) will provide the basis for its rule on regional compliance strategies to assist nonattainment areas to achieve the ozone National Ambient Air Quality Standard (NAAQS) Consisting of representatives of the eastern 17 states, OTAG was a 2-year process of assessing ozone transport, and developing recommendations to mitigate it. OTAG found that ozone transport becomes increasingly important over longer averaging times, indicating the importance of regional controls for attaining the new 8 hour NAAQS. Strategies that OTAG recommended included controls on utilities, non-utility stationary sources, mobile sources, and fuels. However, OTAG's recommendations with respect to utility sources were very broad. Also, OTAG was unable to agree to a specific implementation strategy, opting instead for either a "cap and trade" or a "rate only" approach, depending on a state's discretion and EPA's approval.

OVERVIEW

On July 16, the Administrator of the Environmental Protection Agency (EPA) finalized a new National Ambient Air Quality Standard (NAAQS) for ozone. Accompanying the NAAQS was a white House memorandum discussing the general approach EPA was to take in implementing the new NAAQS.[1] Recognizing ozone as a regional pollution problem, the memorandum focuses on regional compliance strategies for the new NAAQS. In particular,

* Excerpted from CRS Report 97-744 ENR. Updated July 30, 1998.

[1] President Clinton. "Implementation of Revised Air Quality Standards for Ozone and Particulate Matter," Memorandum for the Administrator of the Environmental Protection Agency, July 16, 1997.

the memorandum cites the work of the Ozone Transport Assessment Group (OTAG) as the basis of this approach. As stated therein:

> For the past 2 years the EPA has been working with the 37 most eastern States through the Ozone Transport Assessment Group (OTAG) in the belief that reducing interstate pollution will help *all* areas in the OTAG region attain the NAAQS. ... The OTAG completed its work in June 1997 and forwarded recommendations to the EPA. Based on these recommendations, in September 1997, the EPA will propose a rule requiring States in the OTAG region that are significantly contributing to nonattainment or interfering with maintenance of attainment in downwind States to submit Sips [State Implementation Plans) to reduce their interstate pollution. The EPA will issue the final rule by September 1998.

OTAG represented a cooperative effort between states, the EPA, and numerous stakeholders to address the complex issue of ozone transport. However, opponents of the new ozone NAAQS argue that the OTAG recommendations are far too vague and ambiguous to be cited by EPA as a basis for implementation. This report provides background on the effort, and summarizes OTAG's assessment and recommendations.

BACKGROUND

Interstate transport of air pollutants is not a new problem. However, the 1990 Clean Air Act Amendments (CAAA) provided the Environmental Protection Agency (EPA) and the states with new tools to address the problem. One of those tools is section 176A, a provision that permits the EPA, either on its own or by petition from any state, to establish a transport region to address regional pollution problems contributing to violations of a primary NAAQS. A commission of EPA and state officials is constituted to make recommendations to EPA on appropriate mitigating strategies. Based on the commission's findings and recommendations, EPA is required under section 1 l0(kX5) to notify affected states of inadequacies in their current SIPs and established deadlines (not to exceed 18 months) for submitting necessary revisions.

In 1995, EPA issued a planning guidance memorandum for ozone nonattainment areas classified serious or above.[2] Among its provisions was a call for a 2-year regional assessment of ozone transport and control strategies. The goal was for a consensus between EPA and the affected states on necessary additional regional and other emission reductions to achieve attainment in ozone transport areas. Based on results, states would submit plans by mid-1997 for appropriate local and regional controls to achieve attainment. If no consensus were reached, EPA stated it would use its authority under the Clean Air Act (CAA) to ensure the required reductions are achieved (particularly §126 and/or §110).

In partnership with the Environmental Council of the States (ECO S), OTAG initially met in May, 1995. The group consisted of representatives of the eastern 37 states and other relevant stakeholders, such as industry and environmental groups. The policy arm of OTAG consisted of the state environmental commissioners from the 37 states and the District of Columbia and senior EPA officials. It was this Policy Group that voted on the proposals to be included in recommendations to the states and EPA.

---

[2] Nichols, Mary D. "Ozone Attainment Demonstration," U.S. EPA: Assistant Administrator for Air and Radiation, March 2, 1995.

To develop the necessary assessment data, OTAG created three subgroups and six workgroups to address modeling and assessment issues, strategies and controls, along with outreach and communications. Each subgroup held open working meetings to receive input from various stakeholders. The recommendations of these subgroups were submitted to the Policy Group via joint sessions of the Strategies and Controls and Modeling and Assessment Subgroups. The final OTAG meeting was June 19,1997.

OTAG Assessment Results

OTAG analytical work began with the development of an improved emissions inventory from which extensive modeling of ozone transport could be conducted. From that modeling, two workgroups reached several conclusions with respect to ozone transport and the anticipated benefits from increased nitrogen oxide (NO_X) and volatile organic compound (VOC) control. Among those findings particularly relevant in light of EPA's new 8-hour ozone National Ambient Air Quality Standard (NAAQS) are the following:

- The perceived contribution of ozone transport is strongly dependent on how the ozone "problem" is defined. Local emissions are more important with respect to peak 1-hour concentrations than with respect to lower concentration thresholds and concentrations assessed over longer averaging times (8-hour or seasonal averages), where larger areas and longer distance scales become increasingly important.
- The central portion of the [37-State] OTAG domain is truly unique with respect to ozone and ozone transport. It persistently has elevated ozone levels producing an "ozone pool." Transport in any direction from this region has been implicated with high ozone levels in neighboring areas.
- High ozone levels in the southern portion of the OTAG domain are typically associated with stagnant [pollutant] transport conditions resulting in shorter transport scales than on average. In contrast, high ozone levels in the northern portion of the OTAG domain are more typically associated with higher speed and persistent transport conditions from inside the OTAG domain.
- The proposed 8-hour ozone standard [now finalized] will result in significantly more nonattainment areas across the OTAG domain which will be closer together. This will make ozone transport more critical with respect to nonattainment than it is under the current [now previous] standard.[3]

OTAG Recommendations

Of the 37 states that constituted OTAG, 32 voted in favor of the final recommendations with Alabama, Kentucky, Michigan, Virginia, and West Virginia voting against. Those recommendations covered numerous areas, ranging from utility controls to increased research.

[3] Air Quality Analysis Workgroup. *OTAG Air Quality Analysis Workgroup Results Summary. http://capita.wustl.edu/otag/Reports/ExccSumm/AQASUM3.html*

Size of Transport Region

Which states should comprise the ozone transport region was debated continuously during OTAG deliberations. Particularly, states west of the Mississippi River and some southern states argued that their emissions had de minimus effects on ozone transport. OTAG recommended that these "coarse-grid" states be exempt from OTAG-related controls. These states include North Dakota, South Dakota, Nebraska, Kansas, Oklahoma, Texas, Minnesota, Iowa, Arkansas, Louisiana, Mississippi, and Florida, as well as parts of Maine, New Hampshire, Vermont, New York, Michigan, Wisconsin, Missouri, Alabama, and Georgia. OTAG did recommend that Iowa work with Wisconsin in developing implementation plans for southwestern Wisconsin; that Kansas work with Missouri on implementation plans for Kansas City; and, that Oklahoma, Texas, Arkansas, and Louisiana share modeling results with Missouri.

Utility NO_x Controls

Potential increased NO_x control on utility plants was one of the most contentious issues within OTAG. The final recommendation reflects this contention by being wide enough to cover all points of view. Specifically, OTAG recommended a range in the 26 "fine grid" states between current Clean Air Act controls (i.e., existing requirements) to the lesser of 85% reduction from 1990 rates or a .15 lb/mmBtu performance standard (a significant increase in stringency). It is arguable how such a wide range provides any guidance to EPA on implementing regional controls. Essentially, the range gives EPA a free hand in deciding on appropriate tonnage levels.

OTAG did recommend that any statewide tonnage levels proposed by EPA be revised if additional subregional modeling or air quality analyses justified it. Likewise, OTAG recommended that EPA consider any submitted comments and subregional modeling with respect to proposed statewide tonnage levels prior to finalizing SIP calls.

Non-Utility Stationary Source NO_x Control

OTAG's recommendations for non-utility stationary sources mirrored those for the utility sources. Dividing non-utility facilities into medium and large sources, OTAG recommended that large sources be controlled in proportion to how much utility sources are controlled. If utility sources are required to reduce 55%, large non-utility sources should be required to reduce 55%. This parallel requirement weakens as the utility reductions increase; for an 85% utility source reduction, OTAG recommends only a 70% reduction for large non-utility sources.

Requirements for medium sources are also based on utility source requirement, but considerably less stringent. For 55%-65% utility reductions, no medium non-utility control is recommended. For 75%-85% utility reduction, OTAG recommends that medium non-utility sources be required to install reasonably available control technology (RACT), a determination made by the states based on EPA guidance. As with the utility control recommendations, OTAG urges flexibility if additional modeling and air analyses warrant it.

Federal Emission Control Measures

The federal government has a major role in assisting states achieve compliance, particularly with respect to new stationary source control and mobile source control. OTAG outlines nine specific areas where the federal government should develop, adopt and implement stringent national NO_x and/or VOC control requirements to assist the OTAG region in achieving compliance with the ozone NAAQS. These areas include architectural and industrial maintenance coatings, consumer and commercial products, autobody refinishing, reformulated gasoline, small engines, heavy duty highway engines, heavy duty non-road diesel engines, and locomotive engines.

OTAG also recommended EPA act on some fuel related items. These recommendations included EPA (1) adopting a rule on an appropriate fuel sulfur standard to improve long-term performance of emission control devices, (2) evaluating emission benefits and other effects of cetane adjustments on current diesel engines and adopting standards as appropriate, and (3) considering and potentially adopting new diesel fuel standards by the year 2004 through existing deliberative processes.

Finally, OTAG supported and encouraged adoption of a National Low Emission Vehicle Program.

Specific State Actions

Among the provisions OTAG recommended are that "fine grid" states consider adopting include enhanced inspection and maintenance (I&M) program in urban areas with populations greater than 500,000, and the continued use of reformulated gasoline (RFG) in both mandated and opt-in areas.

Trading Program

Part of the controversy over utility source control involved proposed market-based approaches to implementing such controls. OTAG did considerable work on market mechanisms and found several benefits to these implementation approaches, including: (1) lower compliance costs, (2) incentive for early reductions, (3) incentive for over-control of sources, (4) incentive for innovation, and (5) administrative flexibility.

However, OTAG was unable to agree on a basic structure for a market mechanism. Some states, particularly those in the northeast, advocated a "cap and trade" program. A cap and trade program places a total emission limit (i.e., cap), expressed in tons of pollutants, on specific emission sources in an area. The allowable amount of pollution is then allocated to these sources according to an agreed upon formula. Sources that emit less pollution than their allocation may trade the unused portion of their allocation to sources that exceed their allocation, or bank those unused pollution "credits" for use in a future year. A system like this has been successfully implemented to reduce sulfur dioxide emissions under title IV of the CAA and is being implemented in the Ozone Transport Region (OTR) to reduce NO_x emissions.

Under a cap and trade system, emissions are not permitted to increase over time. Thus, as production increases, new emissions from either new sources (or expanded production at existing sources) has to be offset with increased controls at existing sources. Industry groups have objected to this increasing limitation on their emissions and argued for an alternative that controls the rate of pollution emitted, but not the total amount emitted (i.e., no cap). Under a "rate only" approach, a source would emit as much of a pollutant as it wished, as long as it did not emit them at a rate greater than stipulated under the rate only approach. If a source emitted at below the stipulated rate, it would receive credits amounting the difference between its actual rate and the stipulated rate times its production for a given year. It could trade or bank the resulting credits. If a source exceeded the rate, it could purchase emission credits for the excess amount (based on its total production for a year) from a source that emitted pollution at a rate lower than that agreed upon for that year. Currently, no trading programs are based on this approach.

As was the case with the utility control recommendations, OTAG's final recommendation with respect to a trading program encompasses both points of view. The cap and trade program, called "Track 1," and the rate only program, called "Track 2," are both identified by OTAG as options for states to adopt in implementing utility (and possibly other) control measures. OTAG acknowledged several uncertainties with respect to this recommendation including (1) the need for further work on rate only approaches to implement them with same level of confidence and certainty as the cap and trade system; and (2) how any "cross-track" trading could occur between cap and trade states and rate only states. OTAG did note that EPA would have an integral part in overseeing the implementation and integrity of these approaches.

For its part, OTAG recommends a joint state/EPA workgroup address the issues of implementation of Track I and Track 2, along with identifying key design features that states could select in developing a NO_x trading program. OTAG' believes its work in this area provides a sound basis for the workgroup's task. In addition OTAG suggested the workgroup address other concerns with respect to modeling, market systems, and local control requirements.

CONCLUSIONS

Ozone transport is a complex phenomenon with potential winners and losers. With major stakes involved, it is not surprising that some characterize OTAG's final recommendations as vague and ambiguous, where others see needed flexibility. In any event, OTAG's work assessing ozone transport moves the debate forward and clarifies an issue made more important because of the new 8-hour standard.

OTAG does specify some specific approaches to the problem, particularly with respect to mobile sources and fuel issues. However, with respect to utility control, OTAG leaves open the choice of implementation approaches (such as cap and trade or rate only) and appropriate control levels (such as 0.15 lb./mmBtu for NO_x). Overall, OTAG appears to sanction flexibility, both for the states and for EPA in deciding on how to respond to the ozone transport issue, and how to make necessary reductions.

The White House memorandum outlining the Administration's approach to implementing the new ozone standard states that EPA will propose a rule for the OTAG region in

September 1997. At that time, EPA's interpretation of OTAG's recommendations will be published, with a final rule scheduled for September, 1998. How that process proceeds may reveal the true achievement of OTAG and the future of such cooperative approaches.

Chapter 13

AIR QUALITY STANDARDS: THE DECISIONMAKING PROCESS[*]

John E. Blodgett, Larry B. Parker and James E. McCarthy

INTRODUCTION

The decisions by the Administrator of the Environmental Protection Agency (EPA) in July 1997 to revise the national ambient air quality standards (NAAQS) for ozone and particulate matter (PM) refocused attention on the criteria and the process by which these decisions are made. The new standards have been attacked both as overly stringent and as inadequately protective of health; as ignoring costs and as giving costs too much deference; as going beyond what is scientifically conclusive; and as failing to be sufficiently precautionary.[1]

These controversies, along with other concerns about the Clean Air Act (CAA) and the expiration in 1998 of the authorizations for appropriations in the statute, suggest that the Congress may soon consider amendments to the CAA. If so, the NAAQS decisionmaking process will likely command considerable attention, including how scientific evidence is used and whether costs should be considered in setting NAAQS. Because of discrepancies in views on what transpired - or should have transpired - in EPA's development of the ozone and PM proposals, this report provides background on the processes and procedures for setting and revising NAAQS. It lays out the steps of the decision process and identifies the statutory criteria established by the Act for NAAQS, to aid the reader's understanding of the Act's policy for NAAQS. It does not evaluate the decisionmaking process nor assess EPA's decisions on the ozone and PM NAAQS; nor does it address the several other categories of pollutants (e.g., hazardous air pollutants) or other standards-setting activities under the Act.

---

[*] Excerpted from CRS Report 97-722 ENR. Updated June 24, 1998.

[1] See CRS Report 97-8, Air Quality: Background Analysis of EPA'S 1997 Ozone and Particulate Matter Standards.

BACKGROUND

The process by which EPA sets and revises NAAQS evolved over many years. Initially, federal air quality law focused on supporting state programs, through the conduct of research (a central, federal research program was more efficient than 50 state research efforts) and through technical and financial support (regulation was seen as a state and local matter). By the late 1960s, the federal research effort, located in the National Center for Air Pollution Control, Public Health Service, Department of Health, Education, and Welfare (HEW), included the preparation of "criteria" - a summary of scientific knowledge concerning selected air pollutants - and "guidelines" a summary of control technologies. These criteria and guidelines were to support state programs to control air pollution.

As environmental awareness heightened in the late 1960s, the federal role enlarged. With the Clean Air Act Amendments of 1970, the federal clean air program was moved to the newly created EPA[2] and reconstituted, with the federal government becoming responsible for establishing "national ambient air quality standards" for air pollutants endangering public health and resulting from numerous or diverse mobile or stationary sources. The states remained primarily responsible for developing implementation plans to attain and maintain compliance with those national standards. Major amendments in 1977 and 1990 refined the NAAQS setting and implementation process. The 1977 amendments formally established the Clean Air Scientific Advisory Committee (CASAC) and the 5-year review process. The 1990 amendments increased federal guidance to states in implementing existing NAAQS.

Thus the present process for setting and revising NAAQS consists of the statutory steps incorporated in the CAA over a series of amendments. In addition, several other steps of varying degrees of import have been added by the EPA, by executive orders, and by subsequent regulatory reform enactments by the Congress. The steps are depicted in appendix I; for an example chronology of the steps, see appendix II. The documentary language spelling out those steps, along with explanatory notes, follows.

Note: CAA language is in bold; legislative history language, i.e., from reports on amendments to the CAA, is in *bold Italics;* language from statues other than the CAA is in *italics.*

CRITERIA AIR POLLUTANTS

The pollutants for which NAAQS are set are often called "criteria pollutants." This term reflects the evolution of the process. Before there were NAAQS, the Air Quality Act of 1%7 (a predecessor of the CAA) required the Federal government to list air pollutants and to prepare "criteria" a report summarizing scientific evidence concerning their health effects. The states then set standards. Thus the air pollutants for which NAAQS were later set were originally air pollutants for which the Office of Air Quality in HEW prepared "criteria." The current language specifying the listing of "criteria" air pollutants, as modified by the 1970 and 1977 amendments to the CAA, is as follows..

The Administrator shall publish, and shall from time to time thereafter revise, a list which

[2] EPA was created by Reorganization Plan No.3 in 1970, reprinted as 42 U.S.C. §4321 note; the Agency went into operation December 2 of that year.

includes each air pollutant - (A) emissions of which, in his judgment, cause or contribute to air pollution which may reasonably be anticipated to endanger public **health or welfare; [and] (B) the presence of which in the ambient air results from numerous or diverse mobile or stationary sources**........[CAA §108(a)(1)]

With the Clean Air Act Amendments of 1970, the EPA became responsible for determining national ambient air quality standards for the "criteria pollutants" those of such national scope in release or effect that national standards are appropriate. Also, EPA was to assist states in attaining these standards through grants and technical assistance,[3] and could impose sanctions on states for failures to fulfill their obligations. At present, six pollutants are designated "criteria pollutants" for which NAAQS have been set: particulate matter (PM), ozone (03, a key measure of smog), nitrogen dioxide(NO_2 or, inclusively, nitrogen oxides, NO_x), sulfur oxides (SO_x, or, specifically, S02), carbon monoxide (CO), and lead (Pb).

NATIONAL AMBIENT AIR QUALITY STANDARDS

The Administrator....... shall publish....... regulations prescribing a national primary ambient air quality standard and a national secondary ambient air quality standard for each air pollutant for which air quality criteria have been issued ....[CAA §109(a)(1)(A)]

A NAAQS is a uniform, national standard establishing the maximum permissible concentration of an air pollutant in the ambient air - the "portion of the atmosphere, external to buildings, to which the general public has access."[4] The CAA directs the Administrator to set a NAAQS at a level that provides protection from adverse effects on the public health and welfare. The Act provides for "primary standards" to protect health with a margin of safety and for "secondary standards" to protect welfare.

Primary Standards

National primary ambient air quality standards... shall be ambient air quality standards the attainment and maintenance of which in the judgment of the Administrator, based on such criteria and allowing an adequate margin of safety, are requisite to protect the public health. [CAA § 1 09(b)(1)]

An ambient air quality standard... should be the maximum permissible ambient air level of an air pollution agent or class of such agents (related to a period of time) which will protect the health of any group of the population. [Senate Report No.91-1196, p.10]

---

[3] For example, in conjunction with setting NAAQS, the CAA requires EPA to prepare information on air pollution control technologies:

Simultaneously with the issuance of criteria ..., the Administrator shall, after consultation with appropriate advisory committees and Federal departments and agencies, issue to the States and appropriate air pollution control agencies information on air pollution control technologies, which information shall include data relating to the cost of installation and operation, energy requirements, emission reduction benefits, and environmental impact of the emission control technology. [CAA, § 108(b)(1)]

The preparation of these "guidance documents," as they are known, is not reviewed in this report.

[4] 40 CFR §50.1(e). Air within workplaces is subject to regulation by the Occupational Health and Safety Administration.

As stated in § 109, a NAAQS defines the acceptable concentration of an air pollutant in the ambient air necessary to protect health. As discussed later, costs are not considered in setting primary standards. Areas - defined as air quality control regions - complying with the standard are designated "attainment areas" and areas in which the pollutant exceeds the standard are designated "nonattainment areas." An area can be in attainment for one air pollutant and out of attainment for another.

Although a primary NAAQS has a federally enforceable deadline, a NAAQS does not itself establish what to do when concentrations exceed the standard (nor how to protect clean air in attainment areas). Other provisions of the CAA lay out the process of implementation (specifically, §110). For nonattainment areas, this includes monitoring ambient air quality to determine compliance; requiring states to prepare state implementation plans (SIPs) to bring areas not in compliance into compliance; and requiring states to implement their plans to achieve and maintain the NAAQS by a specified deadline. Federal measures to protect air quality include new source performance standards (§111), which are national standards on specified categories of new sources of selected air pollutants, and mobile source emission standards (§202). Also, the federal government continues to prepare "guidance documents" spelling out available control measures for the NAAQS pollutants. Finally, in attainment areas where the air is cleaner than NAAQS, the CAA establishes a program for new source permitting to "maintain" that clean air quality (Title I, Part C).

"Margin of Safety"

> **.....[A]llowing an adequate margin of safety ....** [CAA §109(b)(1)]
>
> ***In setting such [National ambient air quality] standards the [Administrator] should consider and incorporate not only the results of research summarized in air quality criteria documents, but also the need for margins of safety. Margins of safety are essential to any health related environmental standards if a reasonable degree of protection is to be provided against hazards which research has not yet identified.*** [Senate Committee on Public Works, Report No.91-1196 (1970), pp.9-10]
>
> The phrase, "margin of safety," has been central to several debates.

The phrase seems to imply that NAAQS are based on thresholds that there is a concentration of a pollutant below which adverse health effects do not occur. In this view, the Administrator determines the "no effect" threshold for the pollutant from the scientific evidence in the "criteria document" and then adds a safety factor. From this perspective, the issue is how large the margin of safety should be. In a report on automobile emissions in 1974, a panel of the National Academy of Sciences observed that "the safety factors provided by the air standards are much smaller than is usual in regulating other environmental pollutants such as radioactivity...."[5] This could be interpreted as suggesting the need for tighter standards,[6] thus providing a larger margin of safety. The inability of scientists to find

[5] National Academy of Sciences, Air Quality and Automobile Emission Control, Vol.1 (September 1974), p.6.

[6] U.S. Congress, House, Committee on Interstate and Foreign Commerce, Clean Air Act Amendments of 1977, House Rept. No.95-294, to accompany H.R. 6161 (95th Congress, 1st session) (Washington, D.C.: U.S. Govt. Print. Off., 1977), p.182.

clear thresholds has led some to conclude that costs should be considered in setting NAAQS. A National Academy of Sciences panel observed that

> in no case is there evidence that the threshold levels have a clear physiological meaning, in the sense that there are genuine adverse health effects at and above some level of pollution, but no effects at all below that level. On the contrary, evidence indicates that the amount of health damage varies with the upward and downward variations in the concentration of the pollutant, with no sharp lower limit.[7]

The 1996 reviews of the ozone and PM standards likewise concluded that no threshold of adverse effects could be found for either pollutant. Some argue that if there is no threshold, then there cannot be a margin of safety and as a result the whole NAAQS process of necessity becomes a risk management decision - that is, one in which the Administrator balances risks with costs to decide where to set the standard. Others argue that the lack of a threshold justifies the tightest possible standards.

Another, related debate comes from the view that only in adding a "margin of safety" does the administrator layer a policy judgment onto an objective, scientifically determined NAAQS. Some argue that this judgmental aspect means that the "margin of safety" phrase implicitly endorses the consideration of costs in setting NAAQS; as discussed later, the lead industry sued EPA over lead standards on the basis that the "margin of safety" required EPA to take costs into account in setting NAAQS, but the court ruled that the statute and its legislative history are against that interpretation (Lead Industries Association v. Environmental Protection Agency, 647 F.2d 1 30 (D.C. Cir. 1980)).

The legislative history of the CAA only briefly touches on the "margin of safety" phrase. The Senate Report 90-11% (accompanying legislation that became the Clean Air Amendments of 1970), quoted at the beginning of this section' clearly indicates that the "margin of safety" is designed to protect against the potential for adverse effects to occur at pollutant concentrations below those known to cause harm. Thus, regardless of the existence of a threshold, the margin of safety is a factor the Administrator would consider in making choices involving uncertainties embedded in the definitions of which vulnerable population groups to protect and of what effects are adverse health effects: these issues are discussed later. Similarly, House Report 95-294 (accompanying legislation that became the Clean Air Act Amendments of 1977), after quoting the National Academy of Sciences about the "smaller than usual" safety factor in NAAQS and about the lack of evidence for thresholds, suggested "greater not lesser control of emissions are likely to be needed."' The precautionary premise of the act seems manifest in the phrase "margin of safety" regardless of the existence of thresholds; and the inference that the phrase calls for consideration of costs has been consistently rejected.

Secondary Standards

> **Any national secondary ambient air quality standard... shall specify a level of air quality the attainment and maintenance of which in the judgment of the Administrator, based on such criteria, is requisite to protect the public welfare from any known or anticipated adverse effects associated with presence of such air pollutant in the ambient air.** [CAA §

[7] National Academy of Sciences, *Air Quality and Automobile Emission Control, Vol.1* (September 1974), p.17.

1 09(b)(2)]

> **All language referring to effects on welfare includes, but is not limited to, effects on soils, water, crops, vegetation, man-made materials, animals, wildlife, weather, visibility, and climate, damage to and deterioration of property, and hazards to transportation, as well as effects on economic values and on personal comfort and well-being, whether caused by transformation, conversion, or combination with other air pollutants.** [CAA §302(h)][8]

Secondary NAAQS define the concentration of an air pollutant in the ambient air necessary to protect the "public welfare." Secondary standards are implemented in the same manner as primary NAAQS, with the key difference that there is no federally enforceable specified deadline for attainment. Most secondary NAAQS have been set at the same level as the primary NAAQS.

SETTING NAAQS

Setting and Reviewing NAAQS

The process for setting a NAAQS is a multistage one, and repeats regularly as the CAA requires each NAAQS to be reviewed every 5 years using the same process, to ensure that each NAAQS is based on the most recent scientific information. The CAA is quite specific on certain steps of the process: in particular, on the preparation of a "criteria document" summarizing the scientific information, on the review of that document by an independent scientific committee, on the criteria to be used by the Administrator in deciding on the final standard, and on the procedural process for promulgating the standard. In addition, EPA has administratively added a key step, the preparation of a "staff paper" that summarizes the criteria document and lays out policy options; and Executive Order 12866 requires a Regulatory Impact Analysis (RIA), although the economic analysis it contains is legally irrelevant to the actual decision on the standard. Finally, there are a number of regulatory assessment requirements in law that impinge on the process - but have limited substantive impact on the decision itself These stages are discussed below.

Criteria Document

> **The Administrator shall issue air quality criteria for [each] air pollutant... included... [on the] list....** [CAA §108(a)(2)]

The "criteria document" precedes the NAAQS both in its appearance in air pollution control law and in the process of setting NAAQS. In the early stages of the evolution of Federal air pollution control law, the Federal role focused primarily on research and on providing financial and technical advice to states. This role was exemplified in the requirement that the Public Health Service (which was responsible for Federal air pollution activities before EPA) prepare a "criteria document."

[8] U.S. Congress, House, Committee on Interstate and Foreign Commerce, *Clean Air Act Amendments of 1977,* House Rept. No.95-294, to accompany H.R. 6161 (95th Congress, 1st session) Washington, D.C.: U.S. Govt. Print. Off., 1977), p.182.

"Air quality criteria [documents] are an expression of the scientific knowledge of the relationship between various concentrations of pollutants in the air and their adverse effects on man, animals, vegetation, materials, visibility, and so on.

"Air quality criteria can and should be used in developing air quality standards. Criteria and standards are not synonymous. Air quality criteria are descriptive; that is, likely describe the effects that can be expected to occur whenever and wherever the ambient air level of a pollutant reaches or exceeds a specific figure for a specific time period."
[Dr. Middleton, Director, National Center for Air Pollution Control, Public Health Service, quoted in Senate Committee on Public Works Report 403 on the Air Quality Act of 1967, p.26.]

Air quality criteria for an air pollutant shall accurately reflect the latest scientific knowledge useful in indicating the kind and extent of all identifiable effects on public health or welfare which may be expected from the presence of such pollutant in the ambient air, in varying quantities. [CAA, § 1 08(a)(2)]

In 1970, when the Clean Air Amendments of 1970 established the federal role in setting NAAQS, the "criteria document" became the basic technical underpinning of the standards-setting process. Explicitly, the preparation, review, and use of the "criteria document" was to be objective and scientifically validated. It is scientifically peer reviewed by an advisory committee - established by statute in the 1977 amendments - and by other federal departments and agencies. Much of the review process is open to the public, and the final criteria document is made public:

The issuance of air quality criteria... shall be announced in the Federal Register and copies shall be made available to the general public. [CAA, §108(d))

The criteria document is prepared in the Environmental Criteria and Assessment Office, Office of Health and Environmental Assessment, Office of Research and Development by EPA scientists (with the advice of and review by a scientific advisory committee, described below). Reviewing the scientific literature for all studies relevant to the air pollutant, the preparers consolidate information pertinent to indicating the kinds and magnitudes of effects resulting from the pollutant's presence in ambient air, assess the robustness of the studies, endeavor to resolve inconsistencies, and evaluate findings. Key components of the scientific evidence include epidemiological studies that examine the relationships between ambient pollutant levels and public health and welfare; clinical studies that examine human responses to controlled levels of pollutants, for example in air chambers; and animal studies. Typically, the studies included have undergone peer review and been published in the open literature, but on occasion some other studies, such as preliminary reports on ongoing research, may be included if they meet other standards of scientific reporting.

Criteria documents are major undertakings: the current ozone criteria document contains over 1,500 pages and evaluates nearly 190 scientific studies; the PM criteria document contains 2,400 pages and evaluates some 80 studies. Each study took about one year to draft; review and revisions took another year to come to closure.

Staff Paper

This [staff paper) assessment is intended to help bridge the gap between the scientific review

> contained in the [criteria document] and the judgments required of the Administrator in setting ambient standards for PM. Thus, emphasis is placed on identifying those conclusions and uncertainties in the available scientific literature that the staff believes should be considered in selecting particulate pollutant indicators, forms, averaging times, and levels for the primary (health) and secondary (welfare) standards. [EPA, Particulate Matter Staff Paper (1996), p.I-1]

Based on the criteria document, EPA scientists and policy experts prepare a staff paper. It is developed in the Office of Air Quality Planning and Standards of the Office of Air and Radiation. This document is not required by the CAA; it is an administrative step designed to facilitate the EPA Administrator's decision. It lays Out options for a,NAAQS standard e.g., whether to set a standard, at what level(s) it might be set, and methods for measuring compliance - along with justifications from the criteria document. Like the criteria document, the staff paper is reviewed by the scientific advisory committee.

Clean Air Scientific Advisory Committee

> **The Administrator shall appoint an independent [clean air] scientific review committee [CASAC] composed of seven members including at least one member of the National Academy of Sciences, one physician, and one person representing State air pollution control agencies.**
>
> **....[T]he [clean air scientific advisory] committee... shall... review the criteria published under section 108 and the national primary and secondary ambient air quality standards promulgated under this section and shall recommend to the Administrator any new national ambient air quality standards and revisions of existing criteria and standards as may be appropriate....**
>
> **Such committee shall also (i) advise the Administrator of areas in which additional knowledge is required to appraise the adequacy and basis of existing, new, or revised national ambient air quality standards, (ii) describe the research efforts necessary to provide the required information, (iii) advise the Administrator on the relative contribution to air pollution concentrations of natural as well as anthropogenic activity, and (iv) advise the Administrator of any adverse public health, welfare, social, economic, or energy effects which may result from various strategies for attainment and maintenance of such national ambient air quality standards.** [CAA, §109(d)(2)]

The 1967 Air Quality Act required the Secretary of HEW to consult with "appropriate advisory committees" (along with Federal departments and agencies) when preparing criteria documents. The Secretary established a National Air Quality Criteria Advisory Committee, having a membership broadly representative of industry, universities, conservation interests, and all levels of government. This committee actively participated in the rewriting of the Sulfur Oxides Criteria Document published in 1969, and in the preparation of subsequent criteria documents. This general requirement for consultation was replaced in 1977 by the specific requirements creating CASAC, with its responsibilities for reviewing the scientific basis of the Administrator's decisions on NAAQS. The legislative history concerning the CASAC provision emphasizes its independence:

> ***This committee is intended to assist the Administrator, but it is also intended to have complete independence. This independence will help provide an outside mechanism for***

> ***evaluating whether any pollutant may reasonably be anticipated to endanger public health or environment, for evaluating the scientific and medical data which bear on this question, and for reviewing gaps in the available data and recommending additional needs for research.*** [Committee on Interstate and Foreign Commerce, House Rept. No.95-294 (1977), p.182]

The seven-member CASAC creates a panel to review each NAAQS. This panel consists of the members of CASAC plus consultant members to assure full coverage of the expertises needed to assess fully the issues involved. For the ozone review, a panel of 15 was convened; for the PM review, the panel consisted of 21.

The panel members meet to review each criteria document and staff paper as it is prepared, recommend improvements, and after further meetings and reviews sign off only when they are convinced that each accurately reflects the status of the science. CASAC panel meetings are open to the public.[9] These "closure documents" become part of the record for rulemaking. Thus

> *The independent, scientific review committee's recommendations on these issues will not only aid the Administrator and the Congress, but also the courts in judicial review of any national ambient air quality standard or of the Administrator's failure or refusal to set or revise such a standard with respect to any pollutant.* [Committee on Interstate and Foreign Commerce, House Rept. No.95-294 (1977), p 182]

What the closure letter means is that the CASAC panel members agree that the criteria document and the staff paper provide an adequate scientific basis for regulatory decisionmaking. Using formulaic sentences, a closure letter for a criteria document typically reads:

> At the September 1995 meeting the Panel came to closure on the Criteria Document. It was the consensus of the Panel members that the Criteria Document provides an adequate review of the available scientific data and relevant studies of ozone and related photochemical oxidants. The document is quite comprehensive and will provide an adequate scientific basis for regulatory decisions on ozone and related photochemical oxidants based on available information. [CASAC Closure on the Air Quality Criteria for Ozone and Related Photochemical Oxidants, letter, from Dr. George T. Wolff, Chair, to Honorable Carol M. Browner, Administrator, EPA (November 28, 1995)]

A closure letter for a staff paper typically says:

> It was the consensus of the Panel that although our understanding of the health effects of ozone is far from complete, the document provides an adequate scientific basis for making regulatory decisions regarding a primary ozone standard. [CASAC Closure on the Primary Standard Portion of the Staff Paper for Ozone, letter, from D. George T. Wolff Chair, to Honorable Carol M. Browner, Administrator, EPA November 30, 1995)]

---

[9] CASAC is subject to the Federal Advisory Committee Act (5 U.S.C. App. 2) which governs public accessibility to committee meetings and products.

THE ADMINISTRATOR'S DECISION

....[I]n the judgment of the Administrator.... [CAA, § I 09(b)(1)]

The CAA specifies that the Administrator shall use her "judgment...,based on **[the] criteria [document] and allowing an adequate margin of safety" to** determine the NAAQS **"requisite to protect the public health"** [CAA, §109(b)(1)]. Thus the decision involves weighing of the scientific evidence, collected and analyzed in the criteria document, of the policy options laid out in the staff paper, and of the comments of CASAC following the criteria dictated in the CAA.

Several challenges to the Administrator's judgment have been made under various regulatory provisions of the 1970 Amendments to the Clean Air Act with respect to the criteria EPA should use. A landmark case concerning the appropriate criteria for air quality regulatory judgments *is Ethyl Corp.* v. *EPA.* A 3-judge panel, voting 2-1, invalidated the Administrator's regulation of lead in fuels based on § 211 (regulation of fuels) of the CAA (No. 73-2205 (D.C. Cir. Jan.28, 1975)). The decision hinged on the burden of proof the Administrator had to meet in order to justify the proposed standard, with the majority holding that the Administrator had to show actual harm rather than the threat or risk of harm. Subsequently, the initial decision was vacated and the U.S. Court of Appeals granted a rehearing en banc; the full court upheld the Administrator's regulations on a 5-4 vote, holding that §211 states a merely precautionary standard - i.e., does not require actual harm.

Even though vacated, the initial *Ethyl* decision raised issues affecting all clean air standards setting that were directly addressed in the 1977 Amendments to the CAA.[10] In its report on the bill amending the CAA, the House Committee on Interstate and Foreign Commerce identified six issues raised by the case that ***"required further congressional clarification":***

1. whether the Administrator could "act to prevent harm before it occurs or should he be authorized to regulate an air pollutant only if he finds actual harm has already occurred";
2. whether the Administrator could assess risks or only "make findings of past fact";
3. whether the Administrator could consider the cumulative risk of a pollutant from multiple sources, or only the risk from the single class of sources being regulated;
4. whether different standards of proof apply to different sources;
5. whether "protection of public health" referred only to healthy normal adults or included "susceptible individuals within the exposed population"; and
6. whether a ***"Court's standard of review of informal rulemaking by the Administrator to protect public health be whether he has relied on conclusive or indisputable facts or whether he has reached reasonable conclusions which are rationally justified."***

[10] This discussion, including the quotations, are from U.S. Congress, House, Committee on Interstate and Foreign Commerce, *Clean Air Act Amendments of 1977,* House Rept. No.95-294, to accompany H.R. 6161 (95th Congress, 1" session) (Washington, D.C.: U.S. Govt. Print. Off., 1977), pp.43-51.

To resolve these issues, the Committee added language amending the act, using *"a standardized basis for future rulemaking to protect the public health: the Administrator may regulate a pollutant, emissions of which in his judgment cause or contribute to air pollution which may reasonably be anticipated to endanger public health or welfare.' This same basic formula"* was used in the several standards-setting provisions of the Act, including sections 108 (criteria for NAAQS), 111 (new source performance standards), 112 (hazardous air pollutants), 202 (motor vehicle emission standards), 21 1 (regulation of fuels and fuel additives), and 231 (aircraft emissions), as well as subtitle B of Title I (ozone and stratospheric protection).

In using this language in amending the act, as discussed in the report on the bill, the Committee intended -

1. *To emphasize the preventative or precautionary nature of the act, i.e., to assure that regulatory action can effectively prevent harm before it occurs.......;*
2. *To authorize the Administrator to weigh risks and make reasonable projections of future trends.......;*
3. *To assure consideration of the cumulative impact of all sources of a pollutant in seeing ambient and emission standards, not just the extent of the risk from the emissions from a single source or class of sources of the pollutant;....*
4. *To provide the same standard of proof for regulation of any air pollutant.......;*
5. *To assure that the health of susceptible individuals, as well as healthy adults, will be encompassed in the term 'public health,' regardless of the section of the act under which the Administrator proceeds; and*
6. *To reflect awareness of the uncertainties and limitations in the data which will be available to the Administrator in the foreseeable future to enable him to execute his rulemaking duties under this act, because of the limitations on research resources and the fact that decisionmaking about the risks to public health from air pollution falls on 'the frontiers of scientific and medical knowledge.'*

In short, *"The committee language is intended to emphasize the necessarily judgmental element in the task of predicting future health risks of present action and to confer upon the Administrator the requisite authority to exercise such judgment."*

This is not, however, a blank check: "... *the committee does not intend this language* as *a license for 'crystal ball' speculation. The Administrator's judgment must, of course, remain subject to restraints of reasoned decisionmaking."*

The Evidence

[B]ased on such criteria [document] ... [CAA, § 1 09(b)(I)]

Air quality criteria for an air pollutant shall accurately reflect the latest scientific knowledge useful in indicating the kind and extent of all identifiable effects on public health or welfare which may be expected from the presence of such pollutant in the ambient air, in varying quantities. The criteria for an air pollutant, to the extent practicable, shall include information on

(A) those variable factors (including atmospheric conditions) which of themselves or in

> **combination with other factors may alter the effects on public health or welfare of such air pollutant;**
> **(B) the types of air pollutants which, when present in the atmosphere, may interact with such pollutant to produce an adverse effect on public health or welfare; and**
> **(C) any known or anticipated adverse effects on welfare.** [CAA, § 1 08(a)(2)]

> Each agency shall base its [regulatory] decisions on the best reasonably obtainable scientific, technical, economic, and other information concerning the need for, and consequences of, the intended regulation. [Executive Order 12866]

The evidence for a NAAQS decision ultimately resides in the criteria document, the staff paper, and the CASAC letters of closure, plus materials submitted during the public comment period. Ancillary information is included in the Regulatory Impact Analysis (discussed below). Summaries are included in the proposed and final rules as published in the Federal Register. In setting NAAQS, the generic direction of E.O. 12866 to regulatory agencies to base their "decisions on the best reasonably obtainable scientific, technical, economic, and other information...." is constrained by the CAA, which specifies that health protection is to be the criterion in determining the standard.

The Criteria

Health

Health is the sole criterion for setting the primary NAAQS. The question of whether costs or technical feasibility should be taken into account was litigated in the case of the lead NAAQS, and the decision was unambiguous:

> ... [T]he statute and its legislative history make clear that economic considerations play no part in the promulgation of ambient air quality standards under Section 109.
> Section 109(b) speaks only of protecting the public health and welfare. Nothing in its language suggests that the Administrator is to consider economic or technological feasibility in setting ambient air quality standards.
> The legislative history of the Act also shows the Administrator may not consider economic and technological feasibility in setting air quality standards; the absence of any provision requiring consideration of these factors was no accident; it was the result of a deliberate decision by Congress to subordinate such concerns to the achievement of health goals. *[Lead Industries Association* v. *Environmental Protection Agency,* 647 F.2d 1130 ([).C. Cir. 1980))

Beside the technical information underlying a health standard, the Administrator must consider several crucial policy issues: These include (1) defining whose health the standard is to protect, given that some people will be more susceptible to pollution than others; (2) defining which health effects of pollution are adverse health impacts to be protected against; (3) evaluating exposure and characterizing risk; and (4) considering what factors to take into account in providing "an adequate margin of safety." The statute and legislative history provide some guidance on these questions. As they are key elements of the judgment that the Administrator makes in deciding the standard, these questions are more or less explicitly posed in the staff paper.

Sensitive Populations

For answering the first question, about whose health a primary standard should protect, the Senate Report on the Clean Air Act Amendments of 1970 provided explicit guidance:

> ***In requiring that national ambient air quality standards be established at a level necessary to protect the health of persons, the Committee recognizes that such standards will not necessarily provide for the quality of air required to protect those individuals who are otherwise dependent on a controlled internal environment such as patients in intensive care units or newborn infants in nurseries. However, the Committee emphasizes that included among those persons whose health should be protected by the ambient standard are particularly sensitive citizens such as bronchial asthmatics and emphysematics who in the normal course of daily activity are exposed to the ambient environment In establishing an ambient standard necessary to protect the health of these persons, reference would be made to a representative sample of persons comprising the sensitive group rather than to a single person in such a group.***
> ***Ambient air quality is sufficient to protect the health of such persons whenever there is an absence of adverse effect on the health of a statistically related sample of persons in sensitive groups from exposure to ambient air.*** [Senate Committee on Public Works, Report No.91-1196(1970), p 10)

Also, as discussed earlier, the Committee on Interstate and Foreign Commerce, in its report language discussing amendments relating to the basis for administrative standards, expressly noted its intent to "assure that the health of susceptible individuals, as well as healthy adults, will be encompassed in the term 'public health,' regardless of the section of the act under which the Administrator proceeds."

Adverse Health Effect

The Clean Air Act does not define the adverse health effects against which NAAQS must protect, nor does the legislative history of the CAA contain much discussion of what constitutes an adverse health effect. The National Commission on Air Quality[11] noted proposals that the Act contain greater specificity, but observed:

> any attempt to identify specific types of health effects could inhibit identification of unanticipated effects that should be considered. In addition, most air pollutants affect public health in more than one way. To carry out its responsibilities under the Act, EPA, before making a final decision to set or revise an air quality standard, must have all relevant and reliable scientific information on the full range of possible health effects of a pollutant. [12]

Indications of what constitutes an adverse health effect can be found in the distinction between primary and secondary standards and in the definition of hazardous air pollutants. While a primary NAAQS protects health, a secondary NAAQS protects welfare, which includes, by definition, "personal comfort and well-being." Thus, pollution effects that involve only "personal comfort and well-being" are not health effects that primary standards are directed toward. The definition of a hazardous air pollutant in the 1970 Act is "an air pollutant... ...may cause, or contribute to, an increase in mortality or an increase in serious

[11] The National Commission on Air Quality was created by the Clean Air Act Amendments of 1977, §323, to review the Act and to report to the Congress the effects of the Act, alternative ways of controlling air pollution, and any recommended midcourse corrections.

[12] National Commission on Air Quality, *To Breathe Clean Air* (Washington, D.C.: 1981), p.3.1-2.

irreversible, or incapacitating reversible, illness." Unquestionably, "increase in mortality," irreversible" and "incapacitating" signal health effects. Are bouts of coughing or decreased lung functioning adverse health effects? If they are irreversible or incapacitating, the answer would seem clearly "yes." As a practical matter, if an effect necessitates medical intervention or leads to curtailed activity, such as missing school or work, the answer would seem "yes." The question of whether an effect is an adverse health effect can be difficult to answer, particularly if the effect appears minor but may cumulate over a lifetime.

Exposure Evaluation and Risk Assessment

> ***In order to emphasize the precautionary or preventive purpose of the act (and, therefore, the Administrator's duty to assess risks rather than wait for proof of actual harm), the committee not only retained the concept of endangerment to health; the committee also added the words 'may reasonably be anticipated'....***
>
> ***By its use of the words 'cause or contribute to air pollution" the committee intends to require the Administrator to consider all sources of the contaminant which contribute to air pollution and to consider all sources of exposure to the contaminant -food, water, air, etc. - in determining health risk***
>
> ***Finally, the term 'in the judgment of the Administrator' is intended to modify both the 'cause and contribute to 'phrase and the 'reasonably may be anticipated' phrase*** [Committee on Interstate and Foreign Congress, Clean Air Act Amendments of 1977, Report 95-294 (1970), p. 51.]

As indicated in the previous discussions of the *Ethyl* and *Lead Industries* v. *EPA* cases, and as emphasized in House Report 95-294, the Administrator may promulgate a NAAQS in "reasonable" anticipation of public health endangerment. The practical effect of this criterion is to allow a "weight of evidence" approach to be used in setting a NAAQS; EPA need not prove that adverse health effects have already occurred, nor that the air-borne form of the pollutant solely caused the harm.

Margin of Safety

On the question of what constitutes a "margin of safety," EPA's Staff Paper on Particulate Matter summarizes the situation as follows:

> The U.S. Court of Appeals for the District of Columbia Circuit has held that the requirement for an adequate margin of safety for primary standards was intended to address uncertainties associated with inconclusive scientific and technical information available at the time of standard setting. It was also intended to provide a reasonable degree of protection against hazards that research has not yet identified (Lead Industries Association V. EPA, 647 F.2d 1130, 1154 (D.C. Cir. 1980), *cert denied,* 101 5. Ct. 621(1980); American Petroleum Institute v. Costle, 665 F.2d 1176, 1177 (D.C. Cir. 1981), *cert denied,* 102 S. Ct. 1737(1982)). Both kinds of uncertainties are components of the risk associated with pollution at levels below those at which human health effects can be said to occur with reasonable scientific certainty. Thus, by selecting primary standards that provide an adequate margin of safety, the Administrator is seeking not only to prevent pollution levels that have been demonstrated to be harmful but also to prevent lower pollutant levels that she finds may pose an unacceptable risk of harm, even if the risk is not precisely identified as to nature or degree.
>
> In selecting a margin of safety, the EPA considers such factors as the nature and severity of the health effects involved, the size of the sensitive population(s) at risk, and the kind and degree of the uncertainties that must be addressed. Given that the "margin of safety" requirement by definition only comes into play where no conclusive showing of adverse

effects exists, such factors which involve unknown or only partially quantified risks have their inherent limits as guides to action. The selection of any particular approach to providing an adequate margin of safety is a policy choice left specifically to the Administrator's judgment (*Lead Industries Association* v. *EPA, supra*, 647 F.2d at 1161-62). [EPA, *Particulate Matter Staff Paper* (1996), pp. II-l - II-2]

In effect, the "margin of safety" allows the Administrator to take into account the dimensions of the decision that scientific data cannot resolve.

Costs

As noted above, health being the criterion, costs are not taken into account in setting primary NAAQS. The National Commission on Air Quality explained the principle as follows:

The statutory basis for setting national ambient air quality standards does not take economic factors into account. In the Act, Congress recognized that while the levels of air pollution at which public health is affected generally do not vary among different locations, the costs of meeting a specific standard can vary substantially from area to area, depending upon the severity of the pollution. Thus, if a national air quality standard were based in part on the costs of complying with it, the high costs of meeting the standard in a few heavily polluted areas could result in the standard's being set at a less protective level than is achievable in a reasonable, economic fashion in other areas. The health benefits of good air quality and the economic, social, energy, and other costs of meeting health-based standards can be balanced more effectively and appropriately when control programs are established for particular areas than when national primary standards are set.[13]

The concept of deferring cost considerations until the implementation of standards, rather than incorporating them in standards-setting, arose early in the evolution of the CAA. The Senate Report on the Air Quality Act of 1967 affirmed the primacy of health protection in setting NAAQS:

Considerations of technology and economic feasibility, while important in helping to develop alternative plans and schedules for achieving goals of air quality, should not be used to mitigate against protection of the public health and welfare. [Senate Committee on Public Works, Report No.403 (1967), pp.28-29]

Subsequently, in bringing to the floor the Senate bill that became the Clean Air Act Amendments of 1970, Senator Muskie repeated that language and called it a "warning" that had been ignored by those who sought to compromise health standards because of costs:

That warning... has been on the books of this committee for 3 years, for all to read Contrary 10 this intent, these [cost and feasibility] considerations have been used as arguments to compromise the public health. Therefore the committee has made explicit in this bill what is implicit to standards designed to protect our health. That concept and that philosophy are behind every page of the proposed legislation.

The first responsibility of Congress is not the making of technological or economic Judgments - or even to be limited by what is or appears to be technologically

[13] National Commission on Air Quality, *To Breathe Clean Air* (Washington, D.C.: 1981), p.3.1-2.

> ***feasible. Our responsibility is to establish what the public interest requires to protect the health of persons.***
>
> [Senator Muskie, debate on the National Air Quality Standards Act of 1970 (Congressional Record, September 21, 1970, pp.32901-02)]

It was this explicitness of the statute and of the legislative history behind it that led to the unequivocal judgment in *Lead Industries Association* v. *Environmental Protection Agency* [647 F.2d 1130 (').C. Cir. 1980)], as previously noted. Nevertheless, some still urge that costs can and should be considered in setting NAAQS; such arguments were proffered among the public comments submitted for EPA's 1997 proposed ozone and PM2.5 standards. EPA responded to these comments at some length. [14]

PROMULGATING NAAQS

The procedural steps for promulgating or revising NAAQS is set forth in the CAA itself - not, as is usual elsewhere, in the Administrative Procedure Act[15] Affecting this process to varying degrees are several other statutory requirements affecting regulations, notably the Regulatory Flexibility Act, the Unfunded Mandates Reform Act, and the Small Business Regulatory Enforcement Fairness Act. Also affecting the process are executive mandates, notably Executive Order 12866, Regulatory Planning and Review, and Executive Order 12848, Federal Actions to Address Environmental Justice in Minority Populations and Low-Income Populations.

THE ADMINISTRATIVE RULEMAKING PROCESS

For numerous CAA rulemakings, including the process for promulgating or revising CAA rules, §307 of the CAA modifies the procedures of the Administrative Procedure Act governing rulemaking. The process is codified at 42 U. S.C. 7607(d). In general, the procedures require EPA to establish a "docket" that contains all the crucial elements of the rulemaking, that is open to public inspection, and that represents all the information available for the Administrator's decision. (An additional provision allows evidence not in the docket to be considered in certain cases when the omission was "reasonable.") The key steps of the rulemaking include; (1) notice of the proposed rulemaking, (2) a period available for public comment, (3) promulgation of the rule, which shall include responses to significant comments on the proposal, and (4) an opportunity for judicial review and challenges to the procedural determinations.

---

[14] See, for example, EPA, "Cost Considerations," *National Ambient Air Quality Standards for Ozone; Final Rule,* 62 *Federal Register* 38878-38883 (July 18, 1997).

[15] The Clean Air Act Amendments of 1977 substantially revised these procedures. The reasons for the changes and their intent are discussed at length in U.S. Congress, House, Committee on Interstate and Foreign Commerce, *Clean Air Act Amendments of 1977,* House Rept No 95-294, to accompany H.R. 6161 (95th Congress, lst session) (Washington, D.C.: U S Govt Print Off, 1977), pp.318-325.

Notice of Proposed Rulemaking

> **...[T]he Administrator shall publish, simultaneously with the issuance of such criteria [documentation] and information, proposed national primary and secondary ambient air quality standards ....** [CAA, § I 09(a)(2)]

> **...[N]otice of proposed rulemaking shall be published in the Federal Register,... shall be accompanied by a statement of its basis and purpose and shall specify the period available for public comment**

> **....The notice of proposed rulemaking shall also state the docket number, the location ... of the docket, and the times it will be open to public inspection.** [CAA §307(d)(3)]

As an illustration of the process, on June 12, 1996 EPA published an "Advance Notice of Proposed Rulemaking (ANPR) for National Ambient Air Quality Standards for Ozone and Particulate Matter" (61 *Federal Register* 29719-2925). This ANPR outlined the basis for EPA having to make the decision, identified the key documents, and indicated the major options under consideration. The Proposed Decisions on ozone and PM were released November 27, 1996, and published December 13 (61 *Federal Register 65638-65872).* It laid out the proposed decision, requested public comment generally and on specific options, told where and how to access the docket, and provided for a 60-day public comment period (later extended 21 days, until March 12, 1997).

Public Comment

> **....[A]fter a reasonable time for interested persons to submit written comments thereon....** [CAA, §109(a)(1)(B)]

> **In promulgating a [NAAQS],... (i) the Administrator shall allow any person to submit written comments, data, or documentary information; (ii) the Administrator shall give interested persons an opportunity for the oral presentation of data, views, or arguments....** [CAA, §307(d)(5)]

In the case of the ozone and particulate matter proposed NAAQS, EPA received over 25,000 comments during the public comment period. Also, EPA held 4 public hearings.

Promulgation of the Rule

> **...[After a reasonable time for interested persons to submit written comments thereon (but no later than 90 days after the initial publication of such proposed standards) [the Administrator shall by regulation promulgate such proposed national ambient air quality standards with such modifications as he deems appropriate.** [CAA, §109(a)(1)(B)]

> **(A) The promulgated rule shall be accompanied by (i) a statement of basis and purpose... and (ii) an explanation of the reasons for any major changes in the promulgated rule from the proposed rule. (B) The promulgated rule shall also be accompanied by a response to each of the significant comments, criticisms, and new data submitted in written or oral presentations during the comment period.** [CAA, §307(d)(6)]

The ozone and particulate matter final decisions were signed by the Administrator on July 16, 1997.

Regulatory Impact Assessments[16]

Costs and Benefits - Executive Order 12866

> Each agency shall assess both the costs and the benefits of the intended regulation
>
> For... a significant regulatory action .. the agency shall ... provide... (i) An assessment, including the underlying analysis, of benefits anticipated from the regulatory action (such as, but not limited to, the promotion of the efficient functioning of the economy and private markets, the enhancement of health and safety, the protection of the natural environment, and the elimination or reduction of discrimination or bias) together with, to the extent feasible, a quantification of those benefits; (ii) An assessment, including the underlying analysis, of costs anticipated from the regulatory action (such as, but not limited to, the direct cost both to the government in administering the regulation and to businesses and others in complying with the regulation, and any adverse effects on the efficient functioning of the economy, private markets (including productivity, employment, and competitiveness), health, safety, and the natural environment, together with, to the extent feasible, a quantification of those costs; and (iii) An assessment, including the underlying analysis, of costs and benefits of potentially effective and reasonably feasible alternatives to the planned regulation....
> [Executive Order 12866)

EPA has concluded that NAAQS reviews are "significant" regulatory actions requiring preparation of an RIA. However, EPA also explicitly states that "Because judicial decisions make clear that cost can not be considered in setting NAAQS, the results of the draft RIA have not been considered in developing this proposal" [National Ambient Air Quality Standard for Ozone: Proposed Decision, pp. 157-158].

Unfunded Mandates - Unfunded Mandates Reform Act

> *Unless otherwise prohibited by law, before promulgating any general notice of proposed rulemaking that is likely to result in promulgation of any rule that includes any Federal mandate that may result in the expenditure by State, local, and tribal governments, in the aggregate, or by the private sector, of $100,000,000 or more (adjusted annually for inflation) in any 1 year, and before promulgating any final rule for which a general notice of proposed rulemaking was published, the agency shall prepare a written statement containing-a qualitative and quantitative assessment of the anticipated costs and benefits of the Federal mandate, including the costs and benefits to State, local, and tribal governments or the private sector, as well as the effect of the Federal mandate on health, safety, and the natural environment ..... a description of the extent of the agency's prior consultation with elected representatives ... of the affected State, local, and tribal governments....*
> [Unfunded Mandates Reform Act of 1995, §202(a)]

The 1996 ozone and PM NAAQS proposals are the first NAAQS rulemakings undertaken since enactment of the Unfunded Mandates Reform Act (UNIRA) and the Regulatory

[16] EPA RIA's can be found on http://www.epa.gov/docs/oppe/eaed/eedhmpg.htm the draft RIA's for the 1997 ozone and PM NAAQS can be accessed at http://134.67.104.12/naaqspro/index.htm

Flexibility Act (RFA). While EPA concluded that the proposals were "significant regulatory actions" as defined by E.O. 12866, EPA concluded that the proposals do not trigger the regulatory analysis provisions of UMRA or RFA.

On the issue of unfunded mandates, EPA concludes:

> As indicated previously, EPA cannot consider in setting a NAAQS the economic or technological feasibility of attaining ambient air quality standards, although such factors may be considered to a degree in the development of State plans to implement the standards. Accordingly, EPA has determined that the provision of sections 202, 203, and 205 of the UMRA do not apply to this proposed [NAAQS] decision. The EPA acknowledges, however, that any corresponding revisions to associated State implementation plan requirements and air quality surveillance requirements, 40 CFR part 51 and 40 CFR part 58, respectively, might result in such effects. Accordingly, EPA will address unfunded mandates as appropriate when it proposes any revisions to 40 CFR parts 51 and 58.

[*Proposed Decision: Particulate Matter*, 61 FR 65670 (December 13,1996)]

However, in its PM RIA, EPA does prepare a governmental entities analysis. According to EPA, "This.....is not an unfunded mandates analysis, but provides estimates of the potential budgetary impact of the control measures used in the control strategy-cost analysis affecting State and local government agencies." EPA notes that it "will be useful in guiding future implementation activities...." [PM RIA, p.8-20]

Small Business - Regulatory Flexibility Act

> *Whenever an agency is required... 10 publish general notice of proposed rulemaking for any proposed rule, the agency shall prepare and make available for public comment an initial regulatory flexibility analysis. Such analysis shall describe the impact of the proposed rule on small entities....* [5 U.S.C. §603(a)]
> *When an agency promulgates a final rule..., after being required... to publish a general notice of proposed rulemaking, the agency shall prepare a final regulatory flexibility analysis....* [5 U.S.C. §604(a)]
> *Sections 603 and 604 of this title shall not apply to any proposed or final rule if the head of the agency certifies that the rule will not, if promulgated, have a significant economic impact on a substantial number of small entities.* [5 U.S.C. §605(b)]

On the issue of assessing the impact of regulations on small businesses, EPA said the ozone and PM proposed NAAQS

> will not have a significant economic impact on small entities within the meaning of the RFA. Instead, it will establish a standard of air quality that other Act provisions will call on states (or in case of state default, the federal government), to achieve by adopting implementation plans containing specific control measures for that purpose. In other words, state (or federal) regulations implementing the NAAQS might establish requirements applicable to small entities, but the NAAQS itself would not. For these reasons, the Administrator certifies that this proposed rule will not have a significant economic impact on a substantial number of small entities." [National Ambient Air Quality Standard for Ozone: Proposed Decision, pp.163-164].

EPA nevertheless concedes interest in the "potential impact" of the NAAQS and notes that discussion of those impacts are included in the RIA. In the RIAs, EPA performs a

"Screening Analysis" to "evaluate small entity impacts." This identifies impacts on industries classified by SIC codes.

> *Prior to publication of an initial regulatory flexibility analysis which a covered agency is required to conduct by this chapter –*
>
> *(1) a covered agency shall notify the Chief Counsel for Advocacy of the Small Business Administration and provide the Chief Counsel with information on the potential impacts of the proposed rule on small entities and the type of small entities that might be affected;*
> *(2) not later than 15 days after the date of receipt of the materials described in paragraph (1), the Chief Counsel shall identify individual representative of affected small entities for the purpose of obtaining advice and recommendations from those individuals about the potential impacts of the proposed rule;*
> *(3) the agency shall convene a review panel for such rule consisting wholly of full time Federal employees of the office within the agency responsible for carrying out the proposed rule, the Office of Information and Regulatory Affairs within the Office of Management and Budget, and the Chief Counsel;*
> *(4) the panel shall review any material the agency has prepared in connection with this chapter, including any draft proposed rule, collect advice and recommendations of each individual small entity representative identified by the agency after consultation with the Chief Counsel.*
> *(5) not later than 60 days after the date a covered agency convenes a review panel pursuant to paragraph (3), the review panel shall report on the comments of the small entity representatives and its findings...; and*
> (6) *where appropriate, the agency shall modify the proposed rule, the initial regulatory flexibility analysis or the decision on whether an initial regulatory flexibility analysis is required*
> [Small Business Regulatory Enforcement Fairness Act of 1996, §244]

On the same basis that it decided the Unfunded Mandates and the Regulatory Flexibility Act requirements did not apply when setting NAAQS, EPA concludes "that the small-entity provisions in Section 244 of the Small Business Regulatory Enforcement Fairness Act (SBREFA) do not apply." [National Ambient Air Quality Standard for Ozone: Proposed Decision, p. 164] And similarly, EPA says it "intends to fulfill the spirit of SBREFA on a voluntary basis" by working with the Small Business Administration to hold panel exercises to solicit comments and advice from representatives of small entities.

Other Regulatory Impact Assessments

Besides the regulatory impact assessments required by the statutes discussed above, provisions of the Paperwork Reduction Act may be triggered by reporting requirements. In the ozone and PM proposed rules, EPA indicates that this issue would arise only in implementation and will be addressed in the final RIAs.

Also, Executive Order 12848, Federal Actions To Address Environmental Justice in Minority Populations and Low-Income Populations, requires each federal agency to identify and address, as appropriate, disproportionate adverse health and environmental impacts of programs, policies, and activities on minorities and low-income populations. Again, EPA indicates that analysis of effects on minorities and low-income populations would be appropriately examined in preparation of RIAs in the implementation process.[17]

---

[17] *RIA for Proposed Particulate Matter Ambient Air Quality Standard* (December 1996), p.8-26.

Consultations - Office of Management and Budget; Other Departments and Agencies

> **The drafts of... rules submitted by the Administrator to the Office of Management and Budget for any interagency review process....,all documents accompanying such drafts and all written comments thereon by other agencies and all written responses to such written comments by the Administrator shall be placed in the ...** [CAA, §307(d)(4)(B)(ii)]

> To the extent permitted by law, OMB... shall be the entity that reviews individual regulations.... [Executive Order 12866]

EPA's NAAQS decisions are subject to this OMB review: "In view of its important policy implications, this proposal has been judged to be a 'significant regulatory action' within the meaning of the Executive Order, and EPA has submitted it to OMB for review. Changes made in response to OMB suggestions or recommendations will be documented in the public docket and made available for public inspection...." [National Ambient Air Quality Standard for Ozone: Proposed Decision, p.157]

Likewise, drafts of the proposed and final rule are circulated for review to other departments and agencies. In cases of substantive controversy, as with the ozone and PM25 NAAQS, the issue may go to the White Rouse for final adjudication.

Judicial Review

> **A petition for review of action of the Administrator in promulgating any national primary or secondary ambient air quality standard,... may be filed only in the United States Court of Appeals for the District of Columbia.... Any petition for review... shall be filed within sixty days from the date notice of such promulgation appears in the Federal Register....** [CAA, §307(b)(1)]

EPA's NAAQS rulemaking is subject to several statutory procedural requirements, compliance with which is subject to judicial review. The basic framework is spelled out in the CAA, §307(d), [42 U.S.C. §7607(d)] and details the requirements for public notice and participation in the process. The final rule cannot be based, in whole or part, on any information or data which have not been placed in the rulemaking docket as of the date of its final promulgation. The final rule must be accompanied by a statement of basis and purpose which includes a summary of the factual data upon which the rule is based, the methodology used in obtaining and analyzing the data, the major legal interpretations, and policy considerations underlying the remaking decision. The statement must also contain the agency's response to each of the significant comments, criticisms, and new data submitted in written and oral presentations during the comment period. Courts have also indicated that they will look at the alternatives the agency considered (or believes it should have considered) in assessing necessary compliance. On appeal the court may reverse the rulemaking action if it finds it to be "arbitrary, capricious, an abuse of discretion or otherwise not in accordance with law," a standard by which the courts assess the rule's reasonableness and rationality based on review of the rulemaking record taken as a whole.

> *For any rule subject to this chapter, a small entity that is adversely affected or aggrieved by final agency action is entitled to judicial review of agency compliance with the requirements of sections 601, 604, 605(b) [..if the head of the agency certifies that the rule will not, if promulgated, have a significant economic impact on a substantial number of small entities], 608(b), and 610 in accordance with chapter 7.* (SBREFA, §611(a)(1)]

Separate compliance issues could be raised by the Administrator's certifications that the ozone and PM NAAQS do not directly impact on small entities and therefore EPA does not have to prepare initial or final Regulatory Flexibility Analyses (RFAs) required by the Regulatory Flexibility Act. EPA contends that such impact, if any, can be assessed only at the time of the submission of state implementation plans, or, in default thereof; proposed federal implementation regulations. The Small Business Administration, among others, has objected to that determination, contending that since EPA has ultimate responsibility for implementing the NAAQS it cannot avoid preparing RFA as part of the NAAQS process.[18] The Regulatory Flexibility Act was recently amended by the Small Business Regulatory Enforcement Fairness Act specifically to allow for judicial review of an agency head's certification that a rule will not have a significant economic effect on a substantial number of small entities.

EPA cited the same standards/implementation rationale in deciding that the impact reporting and analysis requirements of the Unfunded Mandates Reform Act of 1995 and the Paperwork Reduction Act were not applicable. Both actions are subject to judicial review.

EPA's decisions that requirements of the Small Business Regulatory Enforcement Fairness Act, the Regulatory Flexibility Act, and the Unfunded Mandates Act do not apply to the ozone and PM NAAQS have been challenged, along with various substantive and procedural aspects of the standards. The outcome of such suits (38 were filed in the U.S. Court of Appeals for the D.C. District during 1997) could have consequences for future NAAQS decision-making processes.

[18] Testimony of Jere W. Glover, Chief Counsel for Advocacy, U.S. Small Business Administration, before the Subcommittees on Government Programs and on Regulatory Reform and Paperwork Reduction, House Committee on Small Business (April 17, 1997), p.9

APPENDIX I. THE NAAQS-SETTING PROCESS

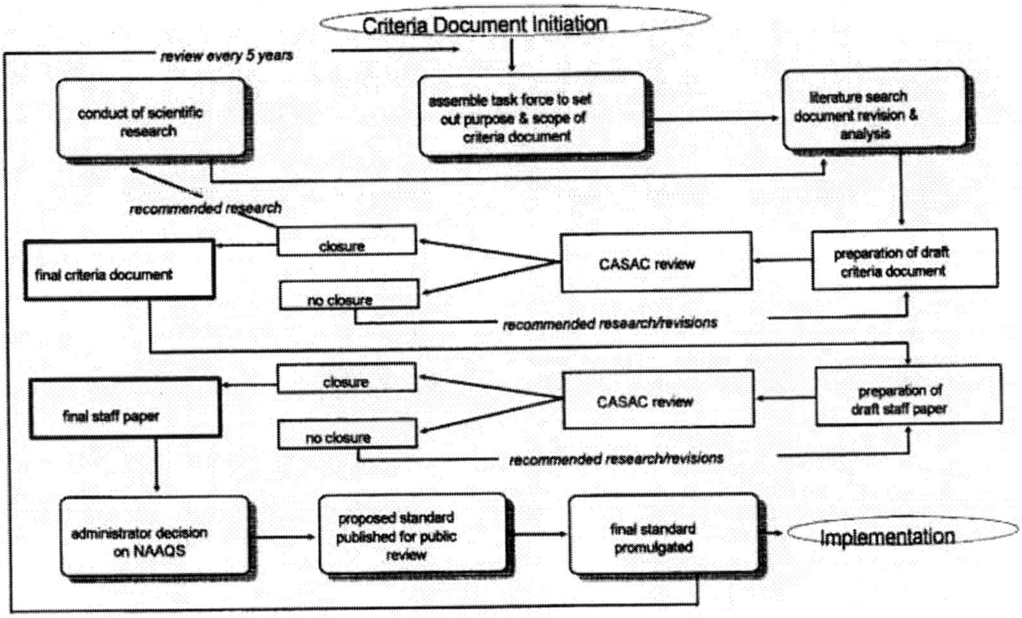

APPENDIX II: SETTING NAAQS A TYPICAL CHRONOLOGY(OZONE)

August 1992	Ozone NAAQS review initiated
Summer/Fall 1993	Workshops
Spring 1994	Draft Criteria Document available to public and CASAC
July 20 and 21, 1994	CASAC meetings to review draft Criteria Document
March 20 and 21, 1995	CASAC meetings to review revised draft of Criteria Document; also to review draft sections of Staff Paper
September 19 and 20, 1995	CASAC closure on Criteria Document; also closure on primary standard section of Staff Paper
November 28, 1995	CASAC closure letter on Criteria Document sent to Administrator
November 30, 1995	CASAC closure letter on primary standard section of Staff Paper sent to Administrator
March 21, 1996	CASAC subpanel meeting on secondary standard section of Staff Paper
April 4, 1996	CASAC closure letter on secondary standard section of Staff Paper sent to Administrator
June 12, 1996	EPA publishes Advance Notice of Proposed Rulemaking(ANPR) for Ozone & PM NAAQS

July 25 and August 8, 1996	Public Meeting on ANPR
November 27, 1996/ December 13 1996	Proposed Decision announced/published in *Federal Register*; public comment period begins
January 14 and 15, 1997	Public Meetings in four cities on proposal
March 12, 1997	End of public comment period on proposal
Winter/Spring 1997	Congressional hearings on the proposed NAAQS
June 25, 1997	President Clinton endorses the proposed ozone & PM NAAQS, with some modifications
Late June 1997	EPA submits proposed final standards to OMB
July 16, 1997	Administrator signs off on final Ozone & PM NAAQS
July 18, 1997	Final NAAQS published in *Federal Register*
July 18, 1997	First suit challenging the final standards filed in the U.S. Court of Appeals for the D.C. District
September 16, 1997	Ozone & PM rules become effective

Chapter 14

MTBE IN GASOLINE: CLEAN AIR AND DRINKING WATER ISSUES*

James E. McCarthy and Mary Tiemann

INTRODUCTION

This chapter provides background information concerning the gasoline additive methyl tertiary butyl ether (MTBE), discusses air and water quality issues associated with it, and reviews options available to congressional and other policy-makers concerned about its continued use. It includes a discussion of legislation in the 108th Congress.

Under the Clean Air Act Amendments of 1990, numerous areas with poor air quality are required to add chemicals called "oxygenates" to gasoline as a means of improving combustion and reducing emissions. The Act has two programs that require the use of oxygenates, but the more significant of the two is the reformulated gasoline (RFG) program, which took effect January 1, 1995.[1] Under the reformulated gasoline program, areas with "severe" or "extreme" ozone pollution (90 counties with a combined population of 64.8 million) must use reformulated gasoline; areas with less severe ozone pollution may opt into the program as well, and many have. In all, portions of 17 states and the District of Columbia use reformulated gasoline (see Table 1 and Figure 1); a little more than 30% of the gasoline sold in the United States is RFG.

The law requires that RFG contain at least 2% oxygen by weight. Refiners can meet this requirement by adding a number of ethers or alcohols, any of which contain oxygen and other

* Excerpted from CRS Report 98-290 ENR. Updated August 11, 2003.

[1] The requirements for reformulated gasoline (RFG), to reduce air toxics and the emissions that contribute to smog formation, are found in Section 211(k) of the Clean Air Act. Separate requirements for oxygenated fuel, to reduce carbon monoxide formation, are contained in Section 211(m). Of the two programs, that for RFG has a much larger impact on the composition of the nation's gasoline, because RFG requirements are in effect year-round and apply to a larger percentage of the country. The Section 211(m) requirements, by contrast are in effect during winter months only and affect a small percentage of the nation's gasoline. Ethanol is the primary oxygenate used in winter oxygenated fuels and MTBE the primary oxygenate used in RFG, although either can be used in both fuels.

elements. Because these substances are not pure oxygen, the amount used to obtain a 2% oxygen level is greater than 2% of the gasoline blend. For example, MTBE is only 19% oxygen and, thus, RFG made with MTBE must contain 11% MTBE by volume to meet the 2% requirement.

By far the most commonly used oxygenate is MTBE. In 1999, 87% of RFG contained MTBE, a number since reduced to about 70%. MTBE has also been used since the late 1970s in non-reformulated gasoline, as an octane enhancer, at lower concentrations. As a result, gasoline with MTBE has been used virtually everywhere in the United States, whether or not an area has been subject to RFG requirements.

Table 1

Areas Using Reformulated Gasoline (as of January 2003)
*Mandatory RFG Areas**
Baltimore, MD
Chicago, IL (and portions of Indiana and Wisconsin)**
Hartford, CT
Houston, TX
Los Angeles, CA
Milwaukee, WI**
New York, NY (and portions of CT and NJ)
Philadelphia, PA (and portions of DE, MD, and NJ)
Sacramento, CA
San Diego, CA
San Joaquin Valley, CA
Southeast Desert, CA
Ventura County, CA
*Opt-In RFG Areas****
Connecticut (entire state)
Dallas / Fort Worth, TX
Delaware (entire state)
District of Columbia
Kentucky portion of Cincinnati metropolitan area
Louisville, KY
Maryland (DC suburbs)
Massachusetts (entire state)
New Hampshire portion of Greater Boston
New Jersey (entire state)
New York (counties near New York City)
Rhode Island (entire state)
St. Louis, MO
Virginia (DC suburbs, Richmond, Norfolk - Virginia Beach - Newport News)

* RFG use required by the Clean Air Act.

** In the Chicago and Milwaukee areas, RFG is made with ethanol rather than MTBE.

*** RFG use required by State Implementation Plan as a means of attaining the ozone air quality standard. These "opt-in" areas may opt out of the program by substituting other control measures achieving the necessary reductions in emissions, but not before January 1, 2004.

Source: U.S. EPA

Federal RFG & Winter OXY/RFG Programs

February, 2000

AIR QUALITY BENEFITS RESULTING FROM MTBE USE

State and local environmental agencies and EPA attribute marked improvements in air quality to the use of fuels containing MTBE and other oxygenates, but the exact role of oxygenates in achieving these improvements is subject to discussion. In Los Angeles, which has had the worst air quality in the country, the use of reformulated gasoline was credited with reducing ground-level ozone by 18% during the 1996 smog season, compared to weather-adjusted data for the same period in 1994 and 1995. Use of RFG also reduced the cancer risk associated with exposure to vehicle emissions bsy 30 to 40%, according to the California EPA, largely because it uses less benzene, a known human carcinogen.[2]

Whether the oxygenates themselves should be given credit for these improvements has been the subject of debate, with the answer depending to some extent on what one assumes would replace the oxygenates if they were removed. Asked to look at the ozone-forming potential of different oxygenates used in reformulated gasoline, a National Academy of Sciences panel concluded that "... the addition of commonly available oxygenates to RFG is likely to have little air-quality impact in terms of ozone reduction."[3] An EPA advisory panel,

---

[2] See "Reformulated Fuels Help Curb Peak Ozone Levels in California," *Daily Environment Report*, November 6, 1996, pp. A-1 and A-2.

[3] Committee on Ozone-Forming Potential of Reformulated Gasoline, National Research Council, *Ozone-Forming Potential of Reformulated Gasoline*, May 1999, p. 5. The NAS study concluded that other characteristics of

by contrast, concluded that the use of oxygenates "appears to contribute to reduction of the use of aromatics with related toxics and other air quality benefits."[4]

Less controversy exists regarding oxygenates' role in reducing carbon monoxide emissions. Both EPA and an interagency group chaired by the White House Office of Science and Technology Policy (OSTP) have reported improvements in carbon monoxide (CO) levels due to the use of oxygenates. According to the June 1997 OSTP report, "analyses of ambient CO measurements in some cities with winter oxygenated gasoline programs find a reduction in ambient CO concentrations of about 10%."[5]

EPA also "believes that the reductions estimated in air quality studies are significant and that these reductions help to protect the public from the adverse health effects associated with high levels of CO in the air."[6] The Agency based its conclusions on both its own analysis and on a report prepared for two industry groups. The latter, using hourly data for more than 300 monitoring sites gathered over a 9-year period, concluded that use of oxygenated fuels was associated with a 14% reduction in ambient CO concentrations.[7]

HEALTH-RELATED QUESTIONS

The improvements in measured air quality have not come without questions. In several cities, residents have complained of a variety of health effects from exposure to MTBE/gasoline exhaust: headaches, dizziness, nausea, sore eyes, and respiratory irritation. Some complaints have centered around the use of MTBE in cold weather, two of the principal areas noting complaints being Alaska and Milwaukee, Wisconsin.

The Interagency Task Force examined these complaints and concluded:

> With regard to exposures ... experienced by the general population and motorists, the limited epidemiological studies and controlled exposure studies conducted to date do not support the contention that MTBE as used in the winter oxygenated fuels program is causing significant increases over background in acute symptoms or illnesses.[8]

Additional research is being conducted by EPA, universities, and others. Under the authority of Section 211 of the Clean Air Act, EPA has requested that refiners conduct a number of

RFG, notably "lowering the Reid Vapor Pressure (RVP) of the fuel, which helps depress evaporative emissions of VOC [volatile organic compounds], and lowering the concentration of sulfur in the fuel, which prevents poisoning of a vehicle's catalytic converter" result in a reduction of about 20% in VOC emissions.

[4] U.S. Environmental Protection Agency, Blue Ribbon Panel on Oxygenates in Gasoline, Executive Summary and Recommendations, July 27, 1999, Appendix A. Available at Internet website: [http://www.epa.gov/otaq/consumer/fuels/oxypanel/blueribb.htm].

[5] Executive Office of the President, National Science and Technology Council, *Interagency Assessment of Oxygenated Fuels*, Washington, D.C., June 1997, p. iv. Referred to hereafter as the OSTP Report.(Executive summary and recommendations are available at Internet website [http://wwwsd.cr.usgs.gov/nawqa/pubs/abstracts/zogorski/ostp.exec.sum.html]). The report expressed some hesitation about its conclusions, particularly regarding the impacts of MTBE in colder weather. It also noted methodological difficulties in identifying statistically significant reductions smaller than 10%, and recommended additional research.

[6] U.S. EPA Response to Interagency Assessment of Oxygenated Fuels, undated, p. 2.

[7] Systems Applications International, Inc., for the Renewable Fuels Association and the Oxygenated Fuels Association, *Regression Modeling of Oxyfuel Effects on Ambient CO Concentrations*, Final Report, January 8, 1997, p. 1.

health effects studies on oxygenated, reformulated, and conventional gasoline, which should provide additional information.

Much discussion has centered on whether MTBE has the potential to cause cancer. Although there are no studies on the carcinogenicity of MTBE in humans, several rodent studies have been done. Based on these animal studies (which looked primarily at inhalation effects), EPA has concluded that MTBE poses a potential for carcinogenicity to humans at high doses; however, because of uncertainties and limitations in the data EPA has been unable to make a confident estimation of risk at low exposure levels.[9] In 1998, the International Agency for Research on Cancer (IARC), the U.S. National Toxicology Program, and California's Carcinogen Identification Committee all determined not to list MTBE as a human carcinogen.

Regarding noncancer effects, another California advisory committee determined that there was not clear scientific evidence to support listing MTBE as a toxic substance affecting human development or reproduction. In reviewing available research on both cancer and noncancer effects, these groups generally noted that research gaps exist, and that the data were particularly limited on health effects associated with MTBE ingestion.

For practical purposes, the interpretation of any health risks associated with the addition of MTBE to gasoline requires a comparison to the health risks associated with conventional gasoline. The Interagency Task Force, EPA, and some environmental groups have all argued that current knowledge suggests that MTBE is a less serious pollutant than the gasoline components it replaced. According to the OSTP report, the cancer risk from exposure to MTBE is "substantially less than that for benzene, a minor constituent of gasoline that is classified as a known human carcinogen; and more than 100 times less than that for 1,3-butadiene, a carcinogenic emission product of incomplete fuel combustion."[10]

WATER QUALITY AND DRINKING WATER ISSUES

A major issue regarding the use of MTBE concerns its detection at low levels in ground water in numerous locations nationwide and at elevated levels in some municipal drinking water wells and reservoirs. MTBE is very soluble and, once released, it moves through soil and into water more rapidly than other chemical compounds present in gasoline. Once in ground water, it is slow to biodegrade and is more persistent than other gasoline-related compounds. In surface water, it dissipates more rapidly: studies show that most of it evaporates from the upper levels of surface water in a few weeks, while it persists longer at greater depths.[11]

---

[8] OSTP Report, p. vi. The report did suggest that "greater attention should be given to the potential for increased symptoms reporting among workers exposed to high concentrations of oxygenated fuels containing MTBE," however.

[9] U.S. Environmental Protection Agency. *Drinking Water Advisory: Consumer Acceptability Advice and Health Effects Analysis on Methyl Tertiary-Butyl Ether (MTBE).* EPA-822-F-97-009, December 1997. p. 1-2. This and other health effects information is available at Internet website: [http://www.epa.gov/OST/drinking/mtbe.html]. (Also, see additional drinking water risk discussion on p. 9 and 10 of this report.)

[10] Ibid., p. vii.

[11] Keller, Arturo, et al., *Health and Environmental Assessment of MTBE,* Report to the Governor and Legislature of the State of California as Sponsored by SB 521, Volume I, Summary and Recommendations, University of California, November 1998. p. 35.

The available data indicate that the primary source of MTBE in ground water has been petroleum releases from leaking underground storage tank (UST) systems. Other significant sources include leaking above ground storage tanks, fuel pipelines, refueling facilities, and accidental spills. The most significant source of MTBE in lakes and reservoirs appears to be exhaust from motorized watercraft, while smaller sources include gasoline spills, runoff, and ground water flow.[12]

Occurrence of MTBE in Drinking Water

Available information on the occurrence of MTBE in public drinking water supplies has increased substantially over the past few years but has been somewhat limited geographically. Although a number of serious contamination incidents have been reported, particularly in California, the available data generally do not indicate a broad presence of MTBE in drinking water supplies at levels of public health concern. However, as monitoring has increased among the states, so has the number of public water systems and private wells showing low-level detections of MTBE.

The most extensive MTBE monitoring data for drinking water are available for California, where testing for MTBE was made mandatory for most public water systems in February 1997. As of April 3, 2002, 2,957 systems had tested 9,905 sources of drinking water. MTBE was detected in 85 (0.9%) of these sources, including 54 (0.6%) of 9,234 ground water sources and 31 (4.6%) of 671 surface water sources. Overall, 53 (1.8%) of the 2,957 public water systems reported detections of MTBE in at least 1 of their drinking water sources, and 13 (0.4%) of the systems reported that a total of 21 (0.2%) sources of water had MTBE concentrations exceeding California's MTBE drinking water standard of 13 micrograms per liter (µg/L).[13]

In 1998, the State of Maine tested nearly 800 public water supplies and 950 randomly selected private wells and found detectable levels of MTBE in 16% of the public water supplies and 15.8% of the private wells. None of the public water supply samples exceeded the state drinking water standard of 35 µg/L, while 1% of private well samples contained MTBE concentrations above the standard. Roughly 94% of public water supply samples showed MTBE levels that were either not detectable or below 1 µg/L; the remaining 6% of samples were between 1 µg/L and 35 µg/L.[14]

Nationwide, the data on the presence of MTBE in drinking water have been more limited. In July 1999, the EPA-appointed Blue Ribbon Panel on Oxygenates in Gasoline reported that between5%and 10% of drinking water supplies tested in high oxygenate use areas show at least detectable amounts of MTBE, and that the vast majority of these detections have been

[12] Keller. p. 33-34.

[13] California Environmental Protection Agency. *MTBE in California Drinking Water,* April 3, 2002. Regular updates are available at Internet website: [http://www.dhs.cahwnet.gov/ps/ddwem/chemicals/MTBE/mtbeindex.htm]. (Micrograms per liter(µg/L) are equivalent to parts per billion (ppb) for fresh water.)

[14] Maine Department of Human Services, Department of Environmental Protection, and Department of Conservation. *The Presence of MTBE and Other Gasoline Compounds in Maine's Drinking Water*, A Preliminary Report. October 13, 1998. 24 p. (Maine was not required to use RFG but had done so voluntarily; the state opted out of the RFG program in October 1998 because of concerns over MTBE contamination of groundwater and drinking water wells.)

well below levels of public health concern, with roughly 1% of detections exceeding 20 µg/L.[15]

More recent federal and state monitoring efforts have been advancing the knowledge about the presence of MTBE in drinking water. Perhaps most notably, the United States Geological Survey (USGS), in cooperation with EPA, recently assessed the occurrence of MTBE and other volatile organic compounds (VOCs) in public water supplies in 10 mid-Atlantic and Northeastern states where MTBE use is common.[16] The study analyzed water from 1,194 randomly selected community water systems. The USGS reported that MTBE was detected in 8.9% of the tested water systems and was strongly associated with areas where reformulated and/or oxygenated (RFG/OXY) fuels are used. Fifteen percent of systems in RFG/OXY areas reported detecting MTBE at concentrations of 1 µg/L or more, while 3% of systems outside of RFG/OXY areas reported such detections. Most MTBE concentrations ranged from 0.5 to 5 µg/L, and less than 1% of the systems reported MTBE at levels equal to or exceeding 20 µg/L, the lower limit of EPA's drinking water advisory.[17]

Occurrence of MTBE in Ambient Ground Water

Looking at ground water generally (not only drinking water wells), the data indicate that low-levels of MTBE are found often. Nationally, the most comprehensive ground water research has been conducted by the USGS through the National Water Quality Assessment Program (NAWQA). USGS data for some 2,743 monitoring, observation, and water supply wells in 42 states (from 1993-1998) showed MTBE present in about5%(145) of the wells, with MTBE levels exceeding 20 µg/L in 0.5% (12) of the wells. In all, MTBE was detected in ground water in 22 of the 42 states. The USGS further evaluated the occurrence data based on whether or not detections occurred in RFG or winter oxyfuel program areas. The researchers reported that low concentrations of MTBE were detected in 21% of ambient ground water samples in high MTBE-use areas and in 2.3% of samples in low or no-MTBE use areas.[18]

MTBE has been detected most frequently in ground water associated with leaking underground storage tank (UST) sites. The California Environmental Protection Agency has estimated that, based on monitoring information available for these sites, MTBE can be expected to be found in shallow, unused ground water at thousands of UST sites in the state, and often at high concentrations (in the parts per million range).[19] Moreover, a 1998 report by

[15] The Blue Ribbon Panel on Oxygenates in Gasoline. Executive Summary and Recommendations. July 27, 1999. Summary and full report are available at Internet website: [http://www.epa.gov/otaq/consumer/fuels/oxypanel/blueribb.htm].

[16] For further information on MTBE research at the USGS, see Internet website: [http://wwwsd.cr.usgs.gov/nawqa/vocns/].

[17] Grady, Stephen J. and George D. Casey. *MTBE and other VOCs in Drinking Water in the Northeast and Mid-Atlantic Region.* Available at Internet website: [http://sd.water.usgs.gov/nawqa/pubs/abstracts/grady/Grady_iccssw_abs.pdf]. MTBE was the second most frequently detected VOC in drinking water, after trihalomethanes (disinfection byproducts) which were detected in 45% of systems tested. Chloroform, the most frequently detected trihalomethane, was found in 39% of systems.

[18] U.S. Geological Survey. Data summary submitted to the EPA Blue Ribbon Panel on the Use of MTBE and Other Oxygenates in Gasoline. January 22, 1999. Available at Internet website: [http://www.epa.gov/otaq/consumer/fuels/oxypanel/blueribb.htm#Presentations].

[19] California Environmental Protection Agency, *MTBE Briefing Paper*, p. 17.

the Lawrence Livermore National Laboratory found that MTBE was not significantly degrading in the monitoring networks for these leaking UST sites.[20]

The picture nationwide may be similar. As of late 2000, 42 states had begun to require testing for MTBE in ground water at leaking UST sites. In a September 2000 survey of state leaking underground storage tank (LUST) programs, 31 states reported that MTBE was found in ground water at 40% or more of gasoline-contaminated sites in their states; 24 states reported MTBE at 60% to 100% of sites.[21]

EPA'S RESPONSES TO MTBE OCCURRENCE IN WATER

Safe Drinking Water Act Initiatives

To address concerns raised by the detection of MTBE in ground water and drinking water supplies, EPA has undertaken a variety of activities. In December 1997, the Agency issued a drinking water advisory for MTBE based on consumer acceptability (for taste and smell). EPA issues drinking water advisories to provide information on contaminants in drinking water that have not been regulated under the Safe Drinking Water Act (SDWA).[22] Advisories are not enforceable, but provide guidance to water suppliers and other interested parties regarding potential health effects or consumer acceptability. While the MTBE advisory is not based on health effects, EPA notes that keeping MTBE levels in the range of 20-40 µg/L or lower for consumer acceptability reasons would also provide a large margin of safety from adverse health effects. Specifically, the advisory states that,

> [c]oncentrations in the range of 20 to 40 µg/L are about 20,000 to 100,000 (or more) times lower than the range of exposure levels in which cancer or noncancer effects were observed in rodent tests. This margin of exposure is in the range of margins of exposure typically provided to protect against cancer effects by the National Primary Drinking Water Standards under the Federal Safe Drinking Water Act. This margin is greater than such standards typically provided to protect against noncancer effects. Thus, protection of the water source from unpleasant taste and odor as recommended will also protect consumers from potential health effects.[23]

Additionally, EPA is taking steps that could lead to the development of an enforceable drinking water standard for MTBE. In February 1998, EPA included MTBE on a list of

---

[20] Happel, Anne, E. H. Beckenbach, and R. U. Halden. *An Evaluation of MTBE Impacts to California Groundwater Resources*. Lawrence Livermore National Laboratory and the University of California, Berkeley. June 11, 1998. p. iv.

[21] NewEngland Interstate Water Pollution Control Commission (NEIWPCC). *Survey of State Experiences with MTBE Contamination at LUST Sites (August 2000)*. Available at Internet website: [http://www.neiwpcc.org/mtbe1.html]. The survey shows that some states began requiring testing at LUST sites in the 1980s (Maine in 1986 and Minnesota in 1987) while others recently began to do so (Kentucky in 2000 and Washington in 2001).

[22] At least 7 states have set health-based drinking water standards for MTBE ranging from 13 parts per billion (ppb) to 240 ppb. (Parts per billion are equivalent to µg/L.) At least 5 states have adopted a secondary standard (based on aesthetic qualities, i.e., taste and odor), ranging from 5 ppb to 70 ppb. At least 32 states have adopted a very wide range of ground water cleanup levels; some are guidelines, some are enforceable, and some vary depending on the use of ground water; some states apply these levels to ground-water cleanup at leaking underground storage tank sites where ground water is used for drinking water.

[23] EPA Drinking Water Advisory, p. 2.

contaminants that are potential candidates for regulation under the Safe Drinking Water Act. Compounds on the contaminant candidate list are categorized as regulatory determination priorities, research priorities, or occurrence priorities. Because of data gaps on MTBE health effects and occurrence, EPA placed MTBE in the category of contaminants for which further occurrence data collection and health effects research are priorities. Thus, while EPA has not selected MTBE for regulation to date, the Agency is pursuing research to fill the existing data gaps so that a regulatory determination may be made.

The Safe Drinking Water Act also directed EPA to publish a rule by August 1999, requiring public water systems to conduct monitoring for a list of unregulated contaminants that may require regulation. EPA included MTBE in this rule and directed large public water systems to begin monitoring for MTBE in January 2001.[24]

The occurrence data generated under the Unregulated Contaminant Monitoring Rule, combined with the results of ongoing health effects studies, are intended to provide information needed by EPA to make a regulatory determination for MTBE. Under SDWA, the next round of regulatory determinations will be made in 2006. EPA typically requires roughly three and one-half years to promulgate a drinking water regulation; thus, the earliest EPA would be expected to issue a drinking water regulation for MTBE is 2010.

Underground Storage Tank Regulation

A key EPA and state contamination prevention effort involves implementing the underground storage tank program established by the 1984 amendments to the Resource Conservation and Recovery Act (RCRA). Under this program, EPA has set operating requirements and technical standards for tank design and installation, leak detection, spill and overfill control, corrective action, and tank closure. As of 1993, all tanks were required to comply with leak detection regulations. Additionally, all tanks installed before December 1988 (when standards for new tanks took effect) were required to be upgraded, replaced or closed by December 22, 1998. Federal and state regulators anticipate that as tank owners and operators comply with these requirements, the number of petroleum and related MTBE leaks from UST systems should decline significantly. Based on reporting by states, EPA estimates that, by the end of fiscal year 2002, 81% of facilities were in "significant operational compliance" with the 1998 release prevention requirements and 74% of facilities were in significant operational compliance with the release detection requirements. However, MTBE has been detected at many leaking tank sites, and this additive is proving more difficult and costly to remediate than conventional gasoline. Moreover, many sites have not been tested for MTBE. A key concern for states is that, as testing increases, it is likely that the number and scope of needed cleanups may increase as well.

In 1986, Congress created a federal response program for cleaning up releases from leaking petroleum USTs through the Superfund Amendments and Reauthorization Act, which amended RCRA Subtitle I. These provisions created the Leaking Underground Storage Tank (LUST) Trust Fund and authorized EPA and states to use the Fund to clean up underground storage tank spills and leaks in cases where tank owners or operators do not clean up sites. EPA and states use the annual Trust Fund appropriation primarily to oversee and enforce corrective actions performed by responsible parties. EPA and states also use Fund monies to conduct corrective actions where no responsible party has been identified, where a responsible party fails to comply with a cleanup order, or in the event of an emergency, and to take cost recovery actions against parties. For FY2003, Congress provided $72.3 million for the program, and the President has requested $72.5 million for FY2004.[25]

In late 2000, EPA launched a new UST fields initiative to address abandoned or idle industrial and commercial properties where redevelopment is hindered by petroleum contamination from abandoned USTs. This program complements EPA's Superfund-related Brownfields program which generally did not cover petroleum contamination until the passage of the new brownfields act (P.L. 107-118). EPA estimates that out of 450,000 brownfields, roughly 100,000 to 200,000 contain abandoned tanks. Under the UST fields initiative, in November 2000, EPA made grants of $100,000 to each of ten communities to clean up abandoned UST sites, with special consideration being given to sites with MTBE contamination. In 2001, the EPA Administrator announced that EPA would provide $100,000 grants from Trust Fund monies to another 40 UST field pilot projects. In P.L. 107-118, signed into law in January 2002, Congress authorized up to $50 million of the annual brownfields appropriation to be used to clean-up petroleum contaminated sites.[26]

Blue Ribbon Panel on Oxygenates in Gasoline

As part of its effort to gather information and focus research, in November 1998, EPA established an independent Blue Ribbon Panel on Oxygenates in Gasoline to review the broad range of issues posed by the use of MTBE and other oxygenates. The panel was established under the auspices of the Clean Air Act Advisory Committee, and its membership reflected a broad range of experts and stakeholders.[27]

The panel was directed to perform the following tasks:

- examine the role of oxygenates in meeting the nation's goal of clean air,
- evaluate the efficiency of each of the available oxygenates in providing clean air benefits and the existence of alternatives,
- assess the behavior of oxygenates in the environment,
- review any known health effects, and
- compare the cost of production and use, and each product's availability.

The panel also was directed to study the causes of ground water and drinking water contamination from motor vehicle fuels, to explore prevention and cleanup technologies for water and soil, and to make recommendations to EPA "on how to ensure public health protection and continued improvement in both air and water quality."

---

[24] 64 *Federal Register* 50555, September 17, 1999. The law requires monitoring by all large public water systems (serving more than 10,000 people) and requires a representative sampling of smaller systems.

[25] For more information on the LUST program and related legislation in the 107th Congress, see CRS Report RS21201, *Leaking Underground Storage Tanks: Program Status and Issues*.

[26] For information on Brownfields, see CRS Issue Brief 10114, *Brownfields and Superfund in the 108th Congress*.

[27] A list of Blue Ribbon Panel members is provided, along with the panel report and related materials, at Internet website: [http://www.epa.gov/oar/caaac/mtbe.html].

Findings and Recommendations

In releasing its recommendations July 27, 1999, the Blue Ribbon Panel stressed that "RFG has provided substantial reductions in the emissions of a number of air pollutants from motor vehicles, most notably volatile organic compounds (precursors of ozone), carbon monoxide, and mobile-source air toxics (benzene, 1,3-butadiene, and others), in most cases resulting in emissions reductions that exceed those required by law."[28]

However, the panel noted water quality problems associated with MTBE releases and made a number of recommendations. Specifically, the panel:

- recommended that Congress act to remove the current Clean Air Act requirement that 2% of RFG, by weight, consist of oxygen, in order to ensure that adequate fuel supplies can be blended in a cost-effective manner while reducing usage of MTBE;
- recommended that the winter oxygenated fuels program be continued;
- agreed broadly that use of MTBE should be reduced substantially (with some members supporting its complete phase out), and that Congress should act to provide clear federal and state authority to regulate and/or eliminate the use of MTBE and other gasoline additives that threaten drinking water supplies;
- recommended that EPA seek mechanisms to ensure that there is no loss of current air quality benefits (i.e., no backsliding); and
- recommended a comprehensive set of improvements to the nation's water protection programs, including over 20 specific actions to enhance Underground Storage Tank, Safe Drinking Water, and private well protection programs.

The panel's numerous water protection recommendations addressed prevention, treatment, and remediation. For example, the panel recommended that EPA work with Congress to determine whether aboveground petroleum storage tanks (which generally are not regulated) should be regulated; work to enhance state and local efforts to protect lakes and reservoirs that serve as drinking water supplies by restricting use of recreational water craft; and accelerate research for developing cost-effective drinking water treatment and remediation technologies.

With regard to the recommendation to reduce substantially the use of MTBE, the panel noted that accomplishing such a major change in gasoline supply without disruptions to fuel supply and price would require up to 4 years lead time if the use of MTBE were eliminated (or less if use was substantially reduced).

The panel also suggested that EPA and others should accelerate ongoing health effects and environmental behavior research of other oxygenates and gasoline components that would likely increase in use in the absence of MTBE.

Then-EPA Administrator Carol Browner concurred with the recommendation of the Blue Ribbon Panel calling for a significant reduction in the use of MTBE. She also stated her

[28] Blue Ribbon Panel on Oxygenates in Gasoline, "Panel Calls for Action to Protect Water Quality While Retaining Benefits from National Clean Burning Gas," press release, July 27, 1999, p. 2. Available at [http://www.epa.gov/oms/consumer/fuels/oxypanel/blueribb.htm]. Regarding dissenting views, one member endorsed the water protection reforms but disagreed with the recommendation to limit the use of MTBE, noting that the panel had not identified any increased public health risk associated with MTBE use in gasoline; another member supported maintaining the existing oxygenate standard for the air quality benefits.

commitment to work with Congress for "a targeted legislative solution that maintains our air quality gains and allows for the reduction of MTBE, while preserving the important role of renewable fuels like ethanol."[29]

On March 20, 2000, she announced that EPA would begin the process of issuing regulations to reduce or phase out use of MTBE (discussed at greater length below in the section on "Current Statutory Authority"). Recognizing that this process could take several years to complete, she renewed her call for congressional action to "amend the Clean Air Act to provide the authority to significantly reduce or eliminate the use of MTBE," to "ensure that air quality gains are not diminished," and to "replace the existing oxygen requirement contained in the Clean Air Act with a renewable fuel standard for all gasoline."[30]

In its few public statements on MTBE, the Bush Administration has not indicated any change in the Clinton Administration's policy, although EPA's effort to regulate MTBE using its existing authority has slowed noticeably. This Administration, like the previous one, would defer to a legislative solution. As one EPA official described it, "If the ethanol and oil industries can come to an agreement, we'll support it."

ALTERNATIVES TO MTBE

The major potential alternatives to MTBE are other oxygenates. This is so both for practical and for regulatory reasons: at present, oxygenates are required by the Clean Air Act, and, they possess several advantages, including high octane and the ability to replace toxic components of conventional gasoline.

Oxygenates that could replace MTBE include ethers, such as ethyl tertiary butyl ether (ETBE), and alcohols, such as ethanol. These other oxygenates may pose health and environmental impacts, but inadequate data make it difficult to reach definite conclusions. EPA's Blue Ribbon Panel concluded:

> The other ethers (e.g., ETBE, TAME, and DIPE) have been less widely used and less widely studied than MTBE. To the extent that they have been studied, they appear to have similar, but not identical, chemical and hydrogeologic characteristics. The Panel recommends accelerated study of the health effects and groundwater characteristics of these compounds before they are allowed to be placed in widespread use.[31]

Ethanol and other alcohols are considered relatively innocuous on their own; they generally do not persist in ground water and are readily biodegraded. However, research suggests that the presence of ethanol in a gasoline plume can extend the spread of benzene and other toxic constituents of gasoline through ground water.[32] This is largely because ethanol is likely to be degraded preferentially by microorganisms that would otherwise feed

[29] Statement by former EPA Administrator Carol Browner on Findings by the EPA's Blue Ribbon MTBE Panel, July 26, 1999, available on the Blue Ribbon Panel home page, previously cited.

[30] U.S. Environmental Protection Agency. "Clinton-Gore Administration Acts to Eliminate MTBE, Boost Ethanol," EPA Headquarters Press Release, March 20, 2000, pp. 7-8.

[31] Blue Ribbon Panel Report, p. 8.

[32] See, for example, "Ethanol-Blended RFG May Cause Small Hike in Gasoline Plume Size," *Mobile Source Report*, December 2, 1999, p. 11, or "Experts Charge Cal/EPA Rushing Approval of Ethanol in RFG," *Inside Cal/EPA*, January 14, 2000, p. 1.

on other chemical components of gasoline including benzene, toluene, ethylbenzene, and xylene (BTEX).

In announcing the phase-out of MTBE in his state, March 25, 1999, California's Governor Davis required three state agencies to conduct additional research on the health and environmental impacts of ethanol, the most likely substitute. In reports approved in January 2000, the agencies concluded that if ethanol were substituted for MTBE, there would be "some benefits in terms of water contamination" and "no substantial effects on public-health impacts of air pollution."[33]

A more recent article, based on the California ethanol review, focused specifically on the relative risks of ground-water contamination by spills of ethanol-blended gasoline, MTBE-blended gasoline, and non-RFG gasoline. The authors concluded that,

> relative to risks associated with standard formulation gasoline, *there is an increase in the risk that wells will be contaminated by RFG using either MTBE or ethanol as an oxygenate.* (Emphasis added.) With ethanol, the risk of contaminating wells decreases after approximately five years. However, the risk continues to grow for MTBE because of the assumption that this chemical is not degraded in the subsurface. The conservative approach used in this analysis, including the low biodegradation rates and assumption that the gasoline source areas are not remediated, results in an overstatement of the risks associated with these additives to gasoline. Nevertheless, the relative trends do favor ethanol when considering risk associated with RFG spills.[34]

The switch from MTBE to ethanol is not without technical problems as well. Ethanol costs substantially more to produce than MTBE; it poses challenges to the gasoline distribution system (it would separate from gasoline if transported long distances by pipeline, so it must be mixed with non-oxygenated gasoline blendstock close to the market in which it is to be sold); and, in the short term, it is unlikely to be available in sufficient quantity to replace MTBE nationwide.[35]

Since late 1997, some refiners have discussed the possibility of making gasoline that meets the performance requirements for RFG without using oxygenates. Tosco and Chevron, two firms with large stakes in the California gasoline market, asked for changes in the rules to allow the sale of RFG not meeting the oxygenate requirement in late 1997. In October 1997, Tosco expressed concern about the growing evidence of the potential for extensive MTBE contamination in asking the California Air Resources Board to "take decisive action" to "begin to move away from MTBE."[36] Chevron, California's largest refiner, followed suit, announcing that it "may be possible to make a cleaner burning gasoline without oxygenates, and still reduce emissions to the same extent achieved with current standards."[37] The

---

[33] California Air Resources Board, Water Resources Control Board, and Office of Environmental Health Hazard Assessment. *Health and Environmental Assessment of the Use of Ethanol as a Fuel Oxygenate.* Report to the California Environmental Policy Council in Response to Executive Order D-5.-99. Dec. 1999.Volume 1, Executive summary. P. 1-22. Report is available at Internet website: [http://www-erd.llnl.gov/ethanol/]).

[34] Powers, Susan, et al. "Will Ethanol-Blended Gasoline Affect Groundwater Quality?" *Environmental Science & Technology*. American Chemical Society. January 1, 2001. p 28A.

[35] For additional information on ethanol, see CRS Report RL30369, *Fuel Ethanol: Background and Public Policy Issues.*

[36] Letter of Duane B. Bordvick, Vice President, Environmental and External Affairs, Tosco, to John D. Dunlap III, Chairman, California Air Resources Board, October 17, 1997.

[37] "ChevronSeeksChanges to Reformulated Gasolines," Press Release, Chevron Corporation Public Affairs Department, December 1, 1997.

company stated its support for legislation allowing it to stop or reduce its use of oxygenates. These statements were supported by the Western States Petroleum Association. The American Petroleum Institute now also supports legislation to remove the RFG oxygenate requirement, which was a key recommendation of the Blue Ribbon Panel.

Affected industries have not been united in seeking authority to replace MTBE, however. The major producers of MTBE have not joined the efforts to promote alternatives, and ethanol producers and agricultural interests (most ethanol is made from corn) are concerned that removing the oxygenate requirement would negatively affect the sales of their products.

CURRENT STATUTORY AUTHORITY TO CONTROL THE USE OF MTBE

Whether EPA has authority to take steps to regulate or ban MTBE use in the absence of specific congressional authorization is a question many have raised as the Agency and Congress consider their responses to MTBE contamination. In theory, if the Agency determines that MTBE poses what it considers a significant threat to air quality, water quality, or human health, it can take action to restrict or ban the substance using existing authority under the Toxic Substances Control Act (TSCA).[38] Until early 2000, based on its public statements, the Agency seemed unlikely to make such a determination. In April 1998 testimony before a House Commerce subcommittee, for example, EPA's then Acting Assistant Administrator for Air and Radiation stated: "One needs to be very cautious about initiating changes to the RFG program that could upset the balance of previous agreements that have led to the significant emissions reductions we are seeing today."[39] Instead, the Agency focused attention on the need to prevent leaks from underground fuel storage tanks, which, it argued, would address the major cause of drinking water contamination by MTBE.

On March 20, 2000, however, former EPA Administrator Browner announced that the Agency would start a regulatory process "aimed at phasing out MTBE,"[40] using Section 6 of TSCA. According to the Agency's press release, the Agency expected to issue a proposed rule to ban or phase down MTBE within 6 months. As the Agency noted, however, a TSCA rulemaking is procedurally burdensome and may take "several years" to complete. To use the authority, the Agency will have to conclude that MTBE poses an unreasonable risk to health or the environment. In the 24 years since TSCA was enacted, the Agency has successfully invoked this authority against fewer than half a dozen classes of chemicals.

The first step in the TSCA rule-making process was the issuance of an Advance Notice of Proposed Rulemaking (ANPRM) on March 24, 2000.[41] The ANPRM solicited the input of interested parties regarding EPA's course of action, including:

- whether some use of MTBE as a gasoline additive should be allowed to continue,

[38] Under the Clean Air Act, EPA has authority to waive the RFG oxygenate requirement if the oxygenate interferes with the attainment of an air quality standard; however, EPA has no authority to waive the requirement for water quality reasons.

[39] Statement of Richard D. Wilson, former Acting Assistant Administrator, Office of Air and Radiation, U.S. EPA, in "Implementation of the Reformulated Gasoline Program in California," Hearing before the Subcommittee on Health and Environment, Committee on Commerce, U.S. House of Representatives, April 22, 1998, Serial No. 105-94, p. 30.

[40] U.S. Environmental Protection Agency. "Clinton-Gore Administration Acts to Eliminate MTBE, Boost Ethanol," EPA Headquarters Press Release, March 20, 2000, p. 2.

- how much lead time would be necessary to allow refiners to eliminate MTBE from RFG or from all fuels without unacceptable impacts on the price or supply of fuel,
- whether EPA should eliminate or cap the use of any other gasoline additives (e.g., other ethers) in addition to MTBE, and
- whether MTBE presents significantly greater risk to public health and/or water quality than alternative gasoline additives.

The Agency also requested additional information regarding releases of gasoline containing MTBE, the extent of contamination of water resources by the substance, remediation technologies, alternatives to MTBE and their potential impacts on health and the environment, and the cost of limiting or phasing out MTBE over various time frames.[42] As of August 2003, the Agency is still preparing a proposed rule, but Agency staff familiar with the process have described it as "basically on hold,"[43] with the Administration deferring to what they think will be congressional action.

In addition to TSCA authority, Section 303 of the Clean Air Act could possibly have been invoked. Section 303 allows the Administrator to seek a restraining order (and temporarily to issue such orders on her own authority) in cases where "a pollution source or combination of sources ... is presenting an imminent and substantial endangerment to public health or welfare, or the environment...." In EPA's assessment, however, studies to date suggest that MTBE is less toxic than certain other gasoline components, such as benzene, so it might be difficult to justify a finding of imminent and substantial endangerment.

LEGISLATION

Legislation that could affect MTBE use has been introduced in every Congress since the 104th, but until last year did not reach the floor of either chamber.[44] In the 107th Congress, however, the Senate included MTBE provisions in its version of H.R. 4, the comprehensive energy bill that it passed April 25, 2002. The House, in its version of H.R. 4, did not pass comprehensive MTBE provisions (the House included only authorization of funds to clean up MTBE leaks in its version of the bill). Senate and House conferees did not reach agreement on the bill.

Relying heavily on the work of the 107th Congress, however, the 108th Congress has moved quickly to address MTBE and related issues. On April 11, 2003, the House passed H.R. 6, a comprehensive energy bill. Title VII of the bill would amend the Clean Air Act to eliminate the requirement that RFG contain 2% oxygen and establish a new requirement that an increasing amount of gasoline contain renewable fuels such as ethanol. The bill would require that 5 billion gallons of renewable fuel be used by 2015 (as compared to 2.1 billion gallons used in 2002). It would also authorize $850 million to be appropriated from the LUST Trust Fund for MTBE cleanup, authorize $750 million to assist the conversion of merchant

---

[41] 65 *FR* 16093, March 24, 2000.

[42] The specific request for information is found on pp. 16106-16107 of the March 24, 2000 Federal Register notice.

[43] Personal communication, U.S. EPA, Office of Transportation and Air Quality, January 6, 2003.

[44] Prior to the 107th Congress, the only legislation that had reached the floor of either house was a Senate amendment to the FY2000 agricultural appropriations bill (S. 1233), offered by Senator Boxer, expressing the sense of the Senate that use of MTBE should be phased out. The Senate adopted the amendment on August 4, 1999.

MTBE production facilities to the production of other fuel additives, preserve the reductions in emissions of toxic substances achieved by the RFG program, and provide a "safe harbor" from product liability lawsuits for producers of MTBE and renewable fuels.[45] The bill would not ban the use of MTBE, however.

On April 9, 2003, the Senate Environment and Public Works Committee ordered S. 791, with similar, but not identical provisions, reported. Language similar to S. 791 is included in the Senate version of H.R. 6, which passed the Senate July 31. Among the differences, the Senate bill *would* ban the use of MTBE in motor fuels 4 years after the date of enactment, except in states that specifically authorize its use. The bill would provide a safe harbor for renewable fuel producers, but, unlike the House bill, would not extend the safe harbor to MTBE. It would authorize the use of $200 million from the LUST Trust Fund for remediating contamination from releases of MTBE and other ether fuel additives (in place of the $850 million authorized by the House), and would authorize an additional $200 million over 6 years for enforcing underground storage tank regulations. The Senate has also passed legislation to strengthen the Leaking Underground Storage Tank cleanup program (S. 195).

Support for waiving the oxygenate requirement is widespread among environmental groups, the petroleum industry, and states; and MTBE use is on the verge of being restricted by many states. A potential obstacle to enacting federal legislation on these subjects has lain among agricultural interests, however. Nearly 10% of the nation's corn crop is used to produce the competing oxygenate, ethanol. If MTBE use is reduced or phased out, but the oxygenate requirement remains in effect, ethanol use would likely soar, increasing demand for corn. Conversely, if the oxygenate requirement is removed by legislation, not only would MTBE use decline, but so, likely, would demand for ethanol.

As a result, Members, Senators, and Governors from corn-growing states have taken a keen interest in MTBE legislation. Unless their interests are addressed, they might pose a potent obstacle to its passage. Reflecting these concerns, both the House and Senate versions of MTBE legislation in the 108th Congress would eliminate the oxygen requirement, but mandate a more than doubling of the use of renewable fuels such as ethanol by 2012 or 2015. This provision appears to have generated broad support for most provisions of the legislation.

California and Other State Initiatives

Among the states, California has arguably been the most active in addressing MTBE issues. Actions taken by the State Legislature and the Governor helped propel the issue to national prominence. Legislation, signed October 8, 1997, required the state to set standards for MTBE in drinking water, and required the University of California to conduct a study of the health effects of MTBE and other oxygenates and risks associated with their use. The UC

[45] The effect of this provision would be to protect anyone in the product chain, from manufacturers down to retailers, from liability for cleanup of MTBE and renewable fuels or for personal injury or property damage based on the nature of the product (a legal approach that has been successfully used in California to require refiners to shoulder liability for MTBE cleanup).With liability for manufacturing and design defects ruled out, plaintiffs would be forced to demonstrate negligence in the handling of such fuels, a more difficult legal standard to meet. As a result, drinking water suppliers widely oppose the safe harbor provision and have expressed concern that it could leave communities paying much of the cost for cleaning up contamination caused by fuels containing MTBE or ethanol. Manufacturers argue that a product liability safe harbor provision is reasonable, given that the fuels are used to meet federal fuel mandates.

report, which was issued in November 1998, recommended a gradual phase-out of MTBE from gasoline in California.[46] Based on the report and on public hearings, Governor Davis issued a finding that "on balance, there is a significant risk to the environment from using MTBE in gasoline in California," and required the state's Energy Commission to develop a timetable for the removal of MTBE from gasoline at the earliest possible date, but not later than December 31, 2002.[47] (This date was amended, in March 2002, to December 31, 2003.) The Governor also required the California Air Resources Board (CARB) to make a formal request to U.S. EPA for a waiver from the requirement to use oxygenates in reformulated gasoline and required three state agencies to conduct additional research on the health and environmental impacts of ethanol, the most likely substitute for MTBE.

The waiver request resulted in months of negotiation between EPA and CARB, with EPA expressing skepticism that it had authority to grant a waiver under the circumstances.[48] The Clean Air Act authorizes waiver of the RFG oxygenate requirement only if the Administrator determines that oxygenates would prevent or interfere with the attainment of a National Ambient Air Quality Standard.[49] More than 2 years later, on June 12, 2001, the Agency finally denied California's request. Without a waiver, gasoline sold in ozone nonattainment areas in the state will be required to contain another oxygenate (most likely, ethanol) when the MTBE ban takes effect, unless Congress acts to change the oxygenate requirement. As of May 2003, about 80% of California's motor fuels were reported to have phased out MTBE in favor of ethanol.

Following California's decision to phase-out MTBE, at least 16 other states (Arizona, Colorado, Connecticut, Illinois, Indiana, Iowa, Kansas, Kentucky, Michigan, Minnesota, Missouri, Nebraska, New York, Ohio, South Dakota, and Washington) have acted to limit or phase out its use. The largest of these, New York, will ban it on January 1, 2004. Maine (which is not required to use RFG, but had chosen to do so) also opted out of the RFG program in October 1998 as a result of concerns over MTBE contamination of ground water, and subsequently substituted a low-volatility gasoline to provide similar reductions in emissions of ozone-forming compounds, without requiring the use of oxygenates.

NAFTA Arbitration

Another MTBE issue that emerged in the wake of California's decision to phase out the use of MTBE in gasoline concerns the applicability of certain provisions in the North American Free Trade Agreement (NAFTA). In June 1999, the Methanex Corporation, a Canadian company that produces methanol in the United States and Canada, notified the U.S. Department of State of its intent to institute an arbitration against the United States under the investor-state dispute provisions of the NAFTA, claiming that the phase-out of MTBE ordered by the Governor of California March 25, 1999 breaches U.S. NAFTA obligations

[46] See Keller, Arturo, et al., *Health & Environmental Assessment of MTBE*, Report to the Governor and Legislature of the State of California As Sponsored by SB 521, November 1998. Available on the web at [http://www.tsrtp.ucdavis.edu/mtberpt/homepage.html].

[47] GovernorGrayDavis, ExecutiveOrderD-5-99. The Executive Order and related materials can be found at: [http://www.governor.ca.gov/state/govsite/gov_homepage.jsp]. (Search "MTBE").

[48] See statements of Robert Perciasepe, former Assistant Administrator for Air and Radiation, U.S. EPA, at the May 6, 1999 House Commerce subcommittee hearing, previously cited, pp. 47-52.

[49] The waiver language is found in Section 211(k)(2)(B).

regarding fair and equitable treatment and expropriation of investments, entitling the company to recover damages which it estimated at $970 million. (Methanol is a major component of MTBE and is Methanex's only product. The California market for MTBE reportedly accounts for roughly 6% of global demand for methanol.)

Chapter 11, Article 1110, of the NAFTA requires the United States, Canada, and Mexico to treat each other's investors and investments in accordance with the principles set out in the Chapter. It also allows these investors to submit to arbitration a claim that a NAFTA party has breached Chapter 11 obligations and to recover damages from any such breach. The NAFTA requires the disputing investor to deliver a written notice of its intent to the NAFTA country involved at least 90 days before the claim is submitted to arbitration under the appropriate international arbitral rules. NAFTA also requires 6 months to elapse "since the events giving rise to a claim" before the investor may proceed with arbitration. Because no settlement was reached within that timeframe, the matter proceeded to arbitration. In August 2002, an arbitration panel ordered Methanex to file a fresh claim more specifically relating the actions of California to the company's manufacture of methanol. Methanex filed a new claim in November 2002.[50]

CONCLUSION

Controversy continues to surround the use of MTBE in gasoline. Research conducted to date suggests that the air quality benefits of its use are substantial. However, increasing detections of MTBE in ground and surface water, and particularly in municipal and private drinking water wells, have raised significant concerns about the use of this oxygenate. Research on MTBE and other oxygenates is ongoing and should provide additional information to help advance the current understanding of MTBE-related health and environmental issues and those of its potential alternatives.

Legislation introduced in Congress initially focused on the limited issue of MTBE use in California, where federal requirements have prevented refiners from adopting a more flexible approach permitted by state regulations. Modifying the federal requirements as they pertain to California has had substantial support among the California congressional delegation. As MTBE has been detected in drinking water wells in other parts of the country, and in surface waters in addition to underground sources, broader legislation has been introduced. These bills emerge in a context of ongoing activities aimed at reducing releases of petroleum, generally, or MTBE, specifically. The effectiveness and sufficiency of these efforts (such as the continued implementation of UST regulations and stricter emissions standards for marine engines), combined with concerns and uncertainties about potential replacements for MTBE, add complexity to the debate. Also, some lawmakers have cautioned against acting precipitously to replace MTBE with other additives without adequate research and consideration of potential adverse consequences. Others view the debate over MTBE as an opportunity to encourage the greater use of ethanol, a competing oxygenate generally made from corn.

[50] For more information, see CRS Report RL31638, *International Investor Protection: "Indirect Expropriation" Claims under NAFTA Chapter 11.*

Developments in the states, particularly California and the Northeast, have driven reconsideration of the petroleum industry's reliance on MTBE as the principal means of meeting RFG requirements. These developments are likely to generate continued congressional interest in the issue. There remain differences over whether to ban MTBE, whether to provide a safe harbor for MTBE producers, and other issues that may delay or prevent enactment of legislation.

Chapter 15

SPORT UTILITY VEHICLES, MINI-VANS, AND LIGHT TRUCKS: AN OVERVIEW OF FUEL ECONOMY AND EMISSIONS STANDARDS*

Brent D. Yacobucci

INTRODUCTION

Sport utility vehicles and other light trucks are regulated through two laws that affect their fuel use and emissions: the Energy Policy and Conservation Act of 1975 (EPCA)[1] and the Clean Air Act (CAA).[2] EPCA first set fuel economy standards for the 1977 model year (MY1977). Since then, all light trucks—including most sport utility vehicles (SUVs) and all mini-vans—have been held to less stringent fuel economy standards than passenger cars. Furthermore, since MY1975, most light trucks have faced less stringent emissions standards under Clean Air Act regulations. The differences in standards reflected the situation prevalent at the time these laws were enacted: first, light trucks were mainly used as commercial and agricultural work vehicles;[3] second, they were a relatively small portion of the U.S. automobile market.

Since then, conditions have changed significantly. Today, light trucks are a larger portion of the total vehicle population, and travel more annual vehicle miles. For example, in 1980, light trucks composed 19.9% of the U.S. new automobile market.[4] By 2001, this figure had increased to 50.8%; SUVs alone accounted for 23.1% of the new vehicle market in 1999, while mini-vans accounted for 5.8%.[5] However, a comparison of market share underestimates this growth and its consequences. While the number of passenger cars sold each year in the

---

* Excerpted from CRS Report RS20298. Updated August 25, 2003.

[1] P. L. 94-163, Section 301; 49 U.S.C. 32902.

[2] 42 U.S.C. 7521.

[3] Heavier-duty uses generally require greater vehicle weight, which tends to increase emissions and decrease fuel economy.

[4] Automotive News. *The 100-Year Almanac and 1996 Market Data Book.* p. 120.

[5] Ward's Communications. *2002 Ward's Automotive Yearbook.* pp. 243-245.

United States has decreased somewhat since 1980, the number of light trucks sold has more than tripled, from 2.2 million in 1980 to 8.7 million in 2001. In 2001, SUV sales alone (4.0 million) nearly doubled total light truck sales for 1980. As a result, the total fuel usage and emissions attributable to these vehicles has increased.

Because of the increased emissions and consumption from light trucks, the Environmental Protection Agency (EPA) has ruled that by MY2009, all light vehicles (including all passenger cars and SUVs, and most vans and pickups—all but the very largest vehicles) will be held to the same emissions standards. Further, the Department of Transportation will increase light truck fuel economy starting in MY2005.

DEFINITION OF "LIGHT TRUCK"

For the purposes of fuel economy standards, DOT defines a light truck as any truck or "truck derivative" with a gross vehicle weight rating (GVWR) of 8,500 pounds or less, and a vehicle curb weight (VCW) of 6,000 pounds or less.[6] (49 CFR 523) SUVs and mini-vans are usually built on truck chassis, or have other truck-like characteristics (e.g. 4 wheel drive and/or flat loading areas), and are therefore classified as light trucks.

EPA uses the same basic definition of a light truck, under the current "Tier 1" emissions standards. However, a distinction is made between different vehicle types (Light-Duty Truck 1 through 4), based on weight and payload.

Finally, some SUVs and pickups are so large (greater than 8,500 lbs. GVWR) that they do not qualify as "light trucks" under the current standards. These include the largest full-sized pickups, passenger vans, and SUVs. However, EPA recently added the classification "medium-duty passenger vehicle" to include heavy passenger vehicles–passenger vans and SUVs up to 10,000 GVWR starting in MY2004. Heavy pickups and cargo vans will not be affected, since their main function is transporting cargo, as opposed to passengers.[7] This change is only for emissions standards, and will not affect fuel economy standards. (See "Congressional Concerns and Activity" in the following section)

FUEL ECONOMY STANDARDS

The Energy Policy and Conservation Act requires Corporate Average Fuel Economy (CAFE) standards for motor vehicles. Under the standards, the average fuel economy of all vehicles of a given class that a manufacturer sells in a model year must be equal to or greater than the standard. These standards were first enacted in response to the desire to reduce petroleum consumption and promote energy security after the Arab oil embargo. The current law sets the CAFE standard for passenger cars at 27.5 miles per gallon (mpg). Furthermore, this law gives DOT the authority to define other classes of vehicles, and set fuel economy standards for those vehicles. Currently, the average fuel economy of light trucks—as they are defined above—is regulated at 20.7 mpg. (49 CFR 533)

---

[6] GVWR: the weight of the vehicle plus the designed maximum load capacity. VCW: the weight of the vehicle with all standard equipment and the fuel tank at nominal capacity.

[7] Environmental Protection Agency (EPA), *Regulatory Announcement: EPA's Program for Cleaner Vehicles and Cleaner Gasoline*. December 21, 1999.

Fuel economy standards for light trucks were steadily tightened through the 1980s and early 1990s, and were completely standardized (for all light truck sub-classes) in MY1996. Passenger car standards have remained unchanged since MY1990. DOT has the regulatory authority to set CAFE standards for a given model year taking into account technological feasibility, economic practicability, other vehicle standards, and the need to conserve energy. Between FY1996 and FY2001, however, Congress expressly prohibited DOT from using any funds to change CAFE standards.[8]

For FY2001, the Senate agreed to accept this CAFE language only if DOT, along with the National Academy of Sciences, was permitted to conduct a study of fuel economy and recommend appropriate CAFE standards (P.L. 106-346). The law required the study take into account the factors outlined in EPCA (feasibility, economics, and energy conservation), as well as vehicle safety and the effects on the auto industry. The panel concluded that light truck fuel economy can be increased without significant cost increases and without harming the auto industry, if given significant lead time.[9] The National Energy Policy Report, produced by the Bush Administration, recommended that DOT review and provide recommendations on increasing CAFE standards, taking into account the NAS study.[10]

New Light Truck CAFE Standards

On April 1, 2003, DOT finalized higher CAFE standards for light trucks starting in MY2005.[11] The new standards would be set at 21.0 mpg for MY 2005, 21.6 mpg for MY2006, and 22.2 mpg for MY2007 (a total increase of 1.5 mpg). In its final rule, DOT predicts that the new standards will save 3.6 billion gallons of gasoline over the life of the vehicles covered. While DOT concluded that the average cost would be only $47 per vehicle to meet the 2007 standards, some auto makers in public comments on the rulemaking contended that the costs would be considerably higher and that the costs would not be shared evenly among manufacturers.[12] Further, manufacturers claimed that DOT's projections for efficiency improvements may not be realized, especially if the upward trend in vehicle size and weight continues.

Also in public comments on the rulemaking, the Environmental Defense and the Union of Concerned Scientists claimed that DOT failed to set the maximum achievable standard, and that the standards could be feasibly set in the range of 23.8 to 24.4 mpg by MY2007. They claimed that DOT made conservative assumptions in developing the proposal and ignored some efficient technologies altogether.[13]

---

[8] For more information on CAFE standards, see CRS Issue Brief IB90122: *Automobile and Light Truck Fuel Economy: Is CAFE Up to Standards?*

[9] National Academy of Sciences, National Research Council, *Effectiveness and Impact of Corporate Average Fuel Economy (CAFE) Standards.* Washington, D.C. 2002.

[10] National Energy Policy Development Group, *National Energy Policy.* May 2001. p. 4-9.

[11] *67 Federal Register 77015-77029.* December 16, 2003.

[12] Pamela Najor, "Big Three Automakers Differ on Ability to Meet Tougher Fuel Economy Standard," *Daily Environment Report.* February 24, 2003. p. A-9.

[13] Pamela Najor, "Environmental Groups Say CAFE Proposal Fails to Achieve Maximum Feasible Standard," *Daily Environment Report.* February 24, 2003. p. A-10.

Other Issues

Even though the CAFE provisions in EPCA were not enacted for the purpose of achieving environmental goals, some environmentalists argue that increased fuel economy would promote better air quality, reduce greenhouse gas emissions, limit dependence on foreign oil, and lower consumer expenditures at the pump. Meanwhile, U.S. automobile industry spokespersons argue that CAFE standards have little or no effect on oil imports, promote more consumption (through lower annual fuel costs), and give an advantage to foreign manufacturers (which produce smaller vehicles). In addition, since EPCA only addresses the need to conserve energy, environmental justifications for tougher standards have been challenged by opponents. Moreover, some consumer groups argue that CAFE standards lead to decreased vehicle safety, because a common method for reducing fuel consumption is to reduce vehicle weight.

Congressional Concerns and Activity

CAFE standards have been a contentious issue in congressional debates. This is especially true as they related to greenhouse gas emissions and the potential for global warming. On April 11, 2003, the House passed its version of H.R. 6, an omnibus energy bill. The House version would not set new CAFE standards, but would require NHTSA to study the feasibility of reducing automobile fuel use by a significant percentage by model year 2012. The Senate version of H.R. 6, passed July 31, 2003, would require NHTSA to issue increased fuel economy standards, but would also freeze CAFE standards for "pickup trucks." However, as there is no existing statutory or regulatory definition for pickup trucks, NHTSA will presumably be required to establish a new technical definition, which may or may not exclude some vans or SUVs.

In addition to the energy bill, a few stand-alone proposals on CAFE have been introduced in the 108th Congress. H.R 1605 (Gilchrest) and S. 255 (Feinstein) would increase light truck CAFE to 27.5 miles per gallon by MY2011. Further, these bills would also amend the definition of "light truck" to include all vehicles up to 10,000 pounds gross weight.

EMISSIONS STANDARDS

Before 1975, all light trucks were classified by EPA as "light duty vehicles", i.e. passenger cars. However, in a case brought by International Harvester, the U.S. Court of Appeals concluded that light trucks should be classified differently, due to the agricultural and commercial nature of their use.[14] Therefore, light trucks were given their own classification and have faced less stringent emissions standards since MY1975. Under the current CAA "Tier 1" standards, light trucks are allowed to emit higher levels of pollution with each heavier weight class. Furthermore, Tier 1 standards for light trucks are generally less stringent than those for passenger cars. Only vehicles in the LDT1 class meet the same

[14] International Harvester Co. v. Ruckelshaus, D.C. Cir. No. 72-1517, Feb. 10, 1973.

standards as passenger cars.[15] Most SUVs and pickups, and all vans, are currently permitted to emit 29% to 47% more carbon monoxide (CO) and 75% to 175% more nitrogen oxides (NO_x) than passenger cars. (40 CFR 86)

Tier 2 Standards

The Clean Air Act gives EPA the authority to set standards for MY2004 and beyond, in order to attain and maintain the National Ambient Air Quality Standards (NAAQS). However, EPA may only tighten emissions standards if the technology to do so is available and cost-effective. On February 10, 2000, EPA finalized new standards for passenger cars and light trucks.[16] Under the new "Tier 2" standards, all light trucks and passenger cars, as well as medium-duty SUVs and passenger vans, will be held to the same emissions standards by MY2009.[17]

Three primary factors influenced EPA's decision: use patterns, market share, and technology. First, according to EPA, today these vehicles "are used like passenger cars and there are more annual miles traveled as a result."[18] Since these vehicles travel more miles per year, emissions attributable to the average vehicle have increased. Second, as was stated above, there are simply more light trucks on the road today. Because of the increased number of these vehicles, the contribution of the emissions from these vehicles to vehicular air pollution has similarly increased. Finally, EPA chose to set such stringent standards for light trucks because new emissions control technology is available to permit cuts in emissions without decreasing weight or power, and without generating excessive costs for automobile manufacturers or consumers.

As in many cases over the past thirty years, EPA is following the lead of California, which has the most stringent automobile emissions standards in the country. In November 1998, California ruled that beginning in MY 2004, all light trucks must meet the same emissions standards as passenger cars.[19]

Issues

The Tier 2 standards have raised three major issues among vehicle manufacturers, oil refiners and environmentalists:

First, because the heaviest vehicles are exempt from the Tier 1 emissions standards (since they are not considered "light trucks"), some environmental groups were concerned that automobile manufacturers would be motivated to "bulk up" their heavier light trucks, placing them in the heavy truck class, to avoid improving their emissions. Therefore, EPA added the class of "medium-duty passenger vehicles" to include these vehicles.

---

[15] These vehicles (compact SUVs and pickups) comprise approximately 17% of the light truck market. Source: EPA. Office of Mobile Sources. *Tier 2/Sulfur Draft Regulatory Impact Analysis*. April 1999. p. V-25.

[16] EPA, *Control of Air Pollution from New Motor Vehicles: Tier 2 Motor Vehicle Emission Standards and Gasoline Sulfur Control Requirements; Final Rule. 65 Federal Register 6698.* February 10, 2000.

[17] CRS Report RS20247: *EPA's Tier 2 Proposal for Stricter Vehicle Emission Standards: A Fact Sheet.*

[18] *64 Federal Register 26035.* May 13, 1999.

[19] CAA allows states to adopt California emissions standards in lieu of the federal standards. In the absence of action by EPA, a number of states planned to adopt the California standards. Auto manufacturers, not wanting to face multiple standards, have been generally supportive of Tier 2.

Second, with existing technology, emissions reductions are dependent on reductions in the sulfur content of gasoline.[20] Therefore, EPA also promulgated more stringent limits on gasoline sulfur content. Oil refiners acknowledged that sulfur does affect emission controls, but claimed that their input was ignored in the rule-making process, and that the timeline for the rule will greatly increase the cost of gasoline, and will push smaller refiners out of the market.[21] Furthermore, they argue that emissions reductions could be achieved with considerably smaller cuts in the sulfur level.[22] Conversely, automobile manufacturers have argued that emissions reductions are unattainable without sulfur reductions, and have expressed support for even greater reductions.

Third, the oil refining industry has questioned whether EPA's justification of the Tier 2 standards relies on the revised NAAQS for ozone and particulate matter, which EPA promulgated in July 1997.[23] These revised standards were remanded to EPA for further consideration by the U.S. Court of Appeals in May 1999.[24] In June 1999, to deter potential legal challenges to the Tier 2 standards, EPA issued a notice that justifies the Tier 2 standards under the previous NAAQS standards for ozone and particulate matter, which still apply to states that have not yet achieved attainment.[25]

Congressional Concerns and Activity

No legislation has been introduced in the 108th Congress on emissions, but vehicle emissions standards are a matter of congressional oversight.

---

[20] Fore more information, see CRS Report RS20163: *Sulfur in Gasoline.*

[21] *Testimony of J. Louis Frank, President, Marathon Ashland Petroleum before the Senate Committee on Environment and Public works*. May 18, 1999.

[22] *Testimony of Clint W. Ensign, Vice President, Government Relations, Sinclair Oil Corporation before the House Committee on Science*. July 21, 1999.

[23] *64 Federal Register 38652-38896.* July 18, 1997.

[24] American Trucking Ass'ns v. EPA, D.C. Cir. No. 94-1440, May 14, 1999.

[25] *64 Federal Register 35112-35119.* June 30, 1999.

INDEX

E

F

G

H

I

P

Q

R

S